Breeding and Protection of Vegetables

Breeding and Protection of Vegetables

Chiranjiv Gandhi

RANDOM

RANDOM PUBLICATIONS
NEW DELHI (INDIA)

Breeding and Protection of Vegetables

ISBN 978-93-5111-364-5

Published in 2014 in India by

RANDOM PUBLICATIONS

4376-A/4B, Gali Murari Lal, Ansari Road
New Delhi-110 002
Phone : +91-11-43580356, +91-11-23289044
e-mail: randomexports@gmail.com, sales@randompublications.com, info@randompublications.com
Reprinted 2021

Type Setting by: Friends Media, Delhi-110089
Digitally Printed at: Replika Press Pvt. Ltd.

Preface

"Vegetable" is a collective term for edible parts of plants. The majority are wild or cultivated annual or biennial herbaceous plants. The consumer values vegetables as nutritional complement providing vitamins and minerals. Vegetables can be divided into leaf vegetables, root vegetables, cabbages, flowering vegetables, bulb vegetables and fruiting vegetables. The breeders continuously adapt their breeding objectives to changing consumer demands, environmental conditions and needs of the trading industry. Some decades ago, lettuce was still a typical spring vegetable. In summer, the then extant lettuce varieties showed a day-length response and bolted. It was only when plant breeding succeeded in developing day-length neutral varieties that lettuce could be cultivated and provided also in summer. Cole crops (brassica oleracea) are another example of what plant breeding can achieve. While wild mustard had been known since Ancient Roman times, today's subspecies have been the work of breeders who crossed them into cauliflower, broccoli, kohlrabi (German turnip), savoy cabbage, white and red cabbage, brussels sprouts, kale. Later on, plant breeders have also concentrated on optimising the cultivation properties, especially by increasing yield performance and picking efficiency. The development of hybrid systems for cole was a major milestone in vegetable breeding. Hybrid varieties showed a hitherto unknown uniformity and also enabled economically viable cultivation on marginal sites. The answer of plant breeders to changing environmental conditions and increased disease or pest pressure is a rich choice of resistant varieties. Resistance in lettuce against downy mildew or clubroot of cauliflower, brussels sprouts or white cabbage do not only improve yield performance, but also the quality of the harvested product.

In response to the strong demand for convenience products, breeders have developed varieties with improved nutritional composition. The post-harvest treatment properties have also been improved, e.g. lettuce varieties that fall apart into equally sized leaves with only one cut. New tomato varieties with firm dense flesh that does not lose its juice when sliced are particularly convenient for sandwiches. With steadily increasing abiotic stress as e.g. drought or lack of nutrients, or biotic stress as e.g. pests, new viroses and bacteria, resistance will continue to be the biggest challenge for plant breeding. In the light of the general trend towards minimized use of plant protection products, new research and breeding methods can help understand genes and their interaction with the environment. They also help to find ways to apply this knowledge in resistance breeding. In the future, however, breeding objectives will also be shaped by changed dietary habits and the needs of the processing industry.

This book will develop in the student an understanding of the principles of this subject. In order to make the subject simple and interesting every topic included in this book has been self-sufficient in itself and has been explained in the light of modern development in a simple and elegant style.

I would like to thank my team for standing beside me throughout my career and writing this book. My special thanks go to "Random Publications" who have published the book.

—Chiranjiv Gandhi

Contents

1

Botanical Description

Tomato plants are vines, initially decumbent, typically growing six feet or more above the ground if supported, although erect bush varieties have been bred, generally three feet tall or shorter. Indeterminate types are "tender" perennials, dying annually in temperate climates (they are originally native to tropical highlands), although they can live up to three years in a greenhouse in some cases. Determinate types are annual in all climates.

Tomato plants are dicots, and grow as a series of branching stems, with a terminal bud at the tip that does the actual growing. When that tip eventually stops growing, whether because of pruning or flowering, lateral buds take over and grow into other, fully functional, vines.

Tomato vines are typically pubescent, meaning covered with fine short hairs. These hairs facilitate the vining process, turning into roots wherever the plant is in contact with the ground and moisture, especially if the vine's connection to its original root has been damaged or severed.

Most tomato plants have compound leaves, and are called regular leaf (RL) plants, but some cultivars have simple leaves known as potato leaf (PL) style because of their resemblance to that close cousin. Of RL plants, there are variations, such as rugose leaves, which are deeply grooved, and variegated, angora leaves, which have additional colours where a genetic mutation causes chlorophyll to be excluded from some portions of the leaves.

The leaves are 10–25 cms (4–10 in) long, odd pinnate, with five to 9 leaflets on petioles, each leaflet up to 8 cms (3 in) long, with a serrated margin; both the stem and leaves are densely glandular-hairy. Their flowers, appearing on the apical meristem, have the anthers fused along the edges, forming a column surrounding the pistil's style. Flowers in domestic cultivars tend to be self-fertilising. The flowers are 1–2 cms (0.4–0.8 in) across, yellow, with five pointed lobes on the corolla; they are borne in a cyme of three to 12 together.

Tomato fruit is classified as a berry. As a true fruit, it develops from the ovary of the plant after fertilisation, its flesh comprising the pericarp walls. The fruit contains hollow spaces full of seeds and moisture, called locular cavities. These vary, among cultivated species, according to type.

Some smaller varieties have two cavities, globe-shaped varieties typically have three to five, beefsteak tomatoes have a great number of smaller cavities, while paste tomatoes have very few, very small cavities. For propagation, the seeds need to come from a mature fruit, and be dried or fermented before germination.

Botanical Classification

In 1753, Linnaeus placed the tomato in the genus *Solanum* (alongside the potato) as *Solanum lycopersicum*. In 1768, Philip Miller moved it to its own genus, naming it *Lycopersicon esculentum*. This name came into wide use, but was in breach of the plant naming rules. Technically, the combination *Lycopersicon lycopersicum* (L.) H.Karst. would be more correct, but this name (published in 1881) has hardly ever been used (except in seed catalogs, which frequently used it and still do).

Genetic evidence has now shown that Linnaeus was correct to put the tomato in the genus *Solanum*, making *Solanum lycopersicum* the correct name.

Both names, however, will probably be found in the literature for some time. Two of the major reasons some still consider the genera separate are the leaf structure (tomato leaves are markedly different from any other *Solanum*), and the biochemistry (many of the alkaloids common to other *Solanum* species are conspicuously absent in the tomato).

Hybrids of tomato and diploid potato can be created in the lab by somatic fusion, and are partially fertile, providing evidence of the close relationship between these species.

Genome Sequencing

An international consortium of researchers from 10 countries, among them researchers from the Boyce Thompson Institute for Plant

Research, began sequencing the tomato genome in 2004, and is creating a database of genomic sequences and information on the tomato and related plants. A prerelease version of the genome was made available in December, 2009. The genomes of its mitochondria and chloroplasts are also being sequenced as part of the project.

The complete genome for the cultivar Heinz 1706 was published on May 31, 2012 in *Nature*. Since many other fruits, like strawberries, apples, melons, and bananas share the same characteristics and genes, researchers stated the published genome could help to improve food quality, food security and reduce costs of all of these fruits.

Breeding

Active breeding programmes are ongoing by individuals, universities, corporations, and organisations. The Tomato Genetic Resource Centre, Germplasm Resources Information Network, AVRDC, and numerous seed banks around the world store seed representing genetic variations of value to modern agriculture.

These seed stocks are available for legitimate breeding and research efforts. While individual breeding efforts can produce useful results, the bulk of tomato breeding work is at universities and major agriculture-related corporations.

These efforts have resulted in significant regionally adapted breeding lines and hybrids, such as the Mountain series from North Carolina. Corporations including Heinz, Monsanto, BHNSeed, Bejoseed, etc., have breeding programmes that attempt to improve production, size, shape, colour, flavour, disease tolerance, pest tolerance, nutritional value, and numerous other traits.

Fruit or Vegetable?

Botanically, a tomato is a fruit: the ovary, together with its seeds, of a flowering plant. However, the tomato has a much lower sugar content than other edible fruits, and is therefore not as sweet. Typically served as part of a salad or main course of a meal, rather than at dessert, it is considered a vegetable for most culinary uses.

One exception is that tomatoes are treated as a fruit in home canning practices: they are acidic enough to be processed in a water bath rather than a pressure cooker as vegetables would require. Tomatoes are not the only food source with this ambiguity: avocadoes, eggplants, cucumbers, and squashes of all kinds (such as zucchini and pumpkins) are all botanically fruits, yet cooked as vegetables. This dispute has led to legal speculation in the United States. In 1887, US tariff laws that imposed a duty on vegetables, but not on fruits, caused the tomato's status to become a matter of legal importance. The US Supreme Court settled this controversy on May 10, 1893, by declaring that the tomato is a vegetable, based on the popular definition that classifies vegetables by use, that they are generally served with dinner and not dessert (*Nix v. Hedden* (149 U.S. 304)).

The holding of this case applies only to the interpretation of the Tariff Act of March 3, 1883, and the court did not purport to reclassify the tomato for botanical or other purposes.

Tomatoes have been designated the state vegetable of New Jersey. Arkansas took both sides by declaring the "South Arkansas Vine Ripe Pink Tomato" to be both the state fruit and the state vegetable in the same law, citing both its culinary and botanical classifications.

In 2009, the state of Ohio passed a law making the tomato the state's official fruit. Tomato juice has been the official beverage of Ohio since 1965. A.W. Livingston, of Reynoldsburg, Ohio, played a large part in popularizing the tomato in the late 19th century; his efforts are commemorated in Reynoldsburg with an annual Tomato Festival.

Names

The scientific species epithet *lycopersicum* means "wolf peach", and comes from German werewolf myths. These said that deadly nightshade was used to summon werewolves, so the tomato's similar, but much larger, fruit was called the "wolf peach" when it arrived in Europe.

The Aztecs called the fruit *xitomatl* (pronounced [ƒiĐÈtomatl]), meaning plump thing with a navel. Other Mesoamerican peoples, including the Nahuas, took the name as *tomatl*, from which some European languages derived the name "tomato".

Tomato Records

The heaviest tomato ever, weighing 3.51 kg (7 lb 12 oz), was of the cultivar 'Delicious', grown by Gordon Graham of Edmond, Oklahoma in 1986. The largest tomato plant grown was of the cultivar 'Sungold' and reached 19.8 m (65 ft) in length, grown by Nutriculture Ltd (UK) of Mawdesley, Lancashire, UK, in 2000.

The massive "tomato tree" growing inside the Walt Disney World Resort's experimental greenhouses in Lake Buena Vista, Florida may be the largest single tomato plant in the world.

The plant has been recognised as a Guinness World Record Holder, with a harvest of more than 32,000 tomatoes and a total weight of 522 kg (1,150 lb). It yields thousands of tomatoes at one time from a single vine. Yong Huang, Epcot's manager of agricultural science, discovered the unique plant in Beijing, China. Huang brought its seeds to Epcot and created the specialised greenhouse for the fruit to grow. The vine grows golf ball-sized tomatoes, which are served at Walt Disney World restaurants.

The world record-setting tomato tree can no longer be seen by guests along the Living With the Land boat ride at Epcot, as the tree developed a disease and was removed in April 2010 after approximately 13 months of life. On August 30, 2007, 40,000 Spaniards gathered in Buñol to throw 115,000 kilograms (250,000 lb) of tomatoes at each other in the yearly Tomatina festival.

Cultural Impact

The town of Buñol, Spain, annually celebrates La Tomatina, a festival centred on an enormous tomato fight. Tomatoes are also a popular "nonlethal" throwing weapon in mass protests, and there was a common tradition of throwing rotten tomatoes at bad performers on a stage during the 19th century; today this is usually referenced as a mere metaphor. Embracing it for this protest connotation, the Dutch Socialist party adopted the tomato as their logo.

The US city of Reynoldsburg, Ohio calls itself "The Birthplace of the Tomato", claiming the first commercial variety of tomato was bred there in the 19th century. Several US states have adopted the tomato as a state fruit or vegetable.

Apiaceae

The Apiaceae (or Umbelliferae), commonly known as carrot or parsley family, is a group of mostly aromatic plants with hollow stems. The family is large, with more than 3,700 species spread across 434 genera, it is the sixteenth largest family of flowering plants. Included in this family are the well known plants: angelica, anise, arracacha, asafoetida, caraway, carrot, celery, centella asiatica, chervil, cicely, coriander/cilantro, cumin, dill, fennel, hemlock, lovage, Queen Anne's Lace, parsley, parsnip, sea holly, and the now extinct silphium.

Description

Most Apiaceae are annual, biennial or perennial herbs (frequently with the leaves aggregated toward the base), though a minority are shrubs or trees. Their leaves are of variable size and alternately arranged, or alternate with the upper leaves becoming nearly opposite. In some taxa the texture is leathery, fleshy, or even rigid, but always with stomata. They are petiolate or perfoliate and more or less sheathing, the blade usually dissected and pinnatifid, but entire in some genera. Most commonly crushing leaves emits a marked smell, aromatic to foetid, but absent in some members. The flowers are nearly always aggregated in terminal umbels, simple or compound, often umbelliform cymes, rarely in heads.

The defining characteristic of this family is the inflorescence: a simple or compound umbel. Flowers across the Apiaceae are fairly uniform and are usually perfect (hermaphroditic) and actinomorphic, but some are andromonoecious, polygamomonoecious, or even dioecious (as in *Acronema*), with a distinct calyx and corolla, but the calyx if often highly reduced, to the point of being undetectable in many species, while the corolla can be white, yellow, pink or purple. The flowers are nearly perfectly pentamerous, with 5 petals, sepals, and stamens.

The androecium contains of 5 stamens, but there is often variation in the functionality of the stamens even within a single inflorescence. Some flowers are functionally staminate (where a pistil may be present but has no ovules capable of being fertilised) while others are functionally pistillate (where stamens are present but their anthers do not produce viable pollen). Pollination of one flower by the pollen of a different flower of the same plant (geitonogamy) is common. The gynoecium consists of two carpels fused into a single, bicarpellate pistil with an inferior ovary. When mature, the fused carpels separate into two mericarps. Stylopodiums secrete nectar, attracting pollinators like flies, mosquitoes, gnats, beetles, moths, and bees.

The fruits are non-fleshy schizocarp of two mericarps, each with a single seed; they separate at maturity and are dispersed by wind. Some fruit segments (like those in "Daucus" spp.) are covered in bristles and spread via external transport. The seeds have an oily endosperm and generally contain large quantities of fatty oils, with the fatty acid petroselinic acid occurring universally throughout the family while rarely being found outside of the Apiaceae.

Systematics

The family is solidly placed within the Apiales order in the APG III classification system. It is closely related to Araliaceae and the boundaries between these families remain unclear. Traditionally groups within the family have been delimited largely based on fruit morphology, and the results from this have not been congruent with the more recent molecular phylogenetic analyses. The subfamilial and tribal classification for the family is currently in a state of flux, with many of the groups being found to be grossly paraphyletic/polyphyletic.

Etymology

The Apiaceae was first described by John Lindley in 1836. The name is derived from the type genus *Apium*, which was originally used by Pliny the Elder circa 50 AD for a celery-like plant. The alternative name for the family, Umbelliferae, derives from the inflorescence being generally in the form of a compound umbel.

The family was one of the first to be recognised as a distinct group in Jacques Daleschamps' 1586 *Historia generalis plantarum*. With Robert Morison's 1672 *Plantarum umbelilliferarum distribution nova* it became the first group of plants for which a systematic study was published.

Cultivation

Many members of this family are cultivated for various purposes. The plant structure includes a tap root, which can be large enough to be useful in food, as with parsnips (*Pastinaca sativa*), carrots (*Daucus carota*), and Hamburg parsley (*Petroselinum crispum*).

Many plants of this group are also adapted to conditions that encourage heavy concentrations of essential oils, and as a result some are flavourful aromatic herbs. Examples are parsley (*Petroselinum crispum*), coriander (*Coriandrum sativum*), and dill (*Anethum graveolens*). The plentiful seeds of the umbers, likewise, are sometimes used in cuisine, as with, coriander (*Coriandrum sativum*), fennel (*Foeniculum vulgare*), cumin (*Cuminum cyminum*), and caraway (*Carum carvi*).

Other notable cultivated Apiaceae include chervil (*Anthriscus cerefolium*), angelica (*Angelica* spp.), celery (*Apium graveolens*), arracacha (*Arracacia xanthorrhiza*), poison hemlock (*Conium maculatum*), sea holly (*Eryngium* spp.), asafoetida (*Ferula asafoetida*), galbanum (*Ferula gummosa*), cicely (*Myrrhis odorata*), anise (*Pimpinella anisum*), lovage (*Levisticum officinale*), and hacquetia (*Hacquetia epipactis*).

Companion Plants

Almost every widely cultivated plant of this group is a companion plant. In large part, this is because the tiny flowers forming the umbels, are perfectly suited for ladybugs, parasitic wasps, and predatory flies, which actually drink nectar when not reproducing. They then will prey upon insect pests on nearby plants. Some of the plants are herbs that produce enough scent to possibly dilute the odours of nearby plants, or the pheromones emitted by insects that find those plants, which would otherwise attract more pests.

Folk Medicine

Many plants in this family, such as wild carrot, have estrogenic properties and have been used as a folk medicine for birth control. Most notable for this use is the extinct giant fennel, silphium. *Conium maculatum* has been used as a sedative and in treatments for arthritis and asthma in addition to its most famous use: as a "humane" method of killing criminals and philosophers.

Daucus carota has been used as a diuretic, as a stimulant, to aid in uterine ailments, and one time as a treatment for jaundice. According to the Greek physician and botanist Dioscorides, "[Parsley] —especially Bastard or Wild Parsley—provokes venery and bodily lust and erection of the parts."

Fennel has been used to treat nausea; species of Ferula have been used to ease constipation and coughs; Cuminum cyminum has been used as a treatment for diarrhea; coriander for halitosis; *Eryngium* spp. to treat genitourinary problems; Centella asiatica to treat syphilis; and *Ferula assa-foetida* has been used as a carminative, antispasmodic, and stimulant.

Chemistry

Polyacetylenes can be found in Apiaceae vegetables like carrot, celery, fennel, parsley and parsnip where they show cytotoxic activities.

Carrot

The carrot (*Daucus carota* subsp. *sativus,*

Etymology: from Late Latin *carôta,* from Greek *karôton,* originally from the Indo-European root *ker-* (horn), due to its horn-like shape) is a root vegetable, usually orange in colour, though purple, red, white, and yellow varieties exist.

It has a crisp texture when fresh. The most commonly eaten part of a carrot is a taproot, although the greens are edible as well. It is a domesticated form of the wild carrot *Daucus carota,* native to Europe and southwestern Asia. The domestic carrot has been selectively bred for its greatly enlarged and more palatable, less woody-textured edible taproot.

Description

It is a biennial plant which grows a rosette of leaves in the spring and summer, while building up the stout taproot, which stores large amounts of sugars for the plant to flower in the second year. The flowering stem grows to about 1 metre (3 ft) tall, with an umbel of white flowers that produce a fruit called a mericarp by botanists, which is a type of schizocarp.

Methods of Consumption and Uses

Carrots can be eaten in a variety of ways. Only 3% of the â-carotene in raw carrots is released during digestion: this can be improved to 39% by pulping, cooking and adding cooking oil. Alternatively they may be chopped and boiled, fried or steamed, and cooked in soups and stews, as well as baby and pet foods. A well known dish is *carrots julienne*. The greens are edible as a leaf vegetable, but are rarely eaten by humans. Together with onion and celery, carrots are one of the primary vegetables used in a *mirepoix* to make various broths.

In India carrots are used in a variety of ways, as salads or as vegetables added to spicy rice or daal dishes. The most popular variation in north India is the Gaajar Kaa Halwaa carrot dessert, which has carrots grated and cooked in milk until the whole mixture is solid, after which nuts and butter are added.

Carrot salads are usually made with grated carrots in western parts with a seasoning of mustard seeds and green chillies popped in hot oil, while adding carrots to rice usually is in julienne shape.

The variety of carrot found in north India is rare everywhere except in Central Asia and other contiguous regions, and is now growing in popularity in larger cosmopolitan cities in South India.

The north Indian carrot is pink-red comparable to plum or raspberry or deep red apple in colour (without a touch of yellow or blue) while most other carrot varieties in the world vary from orange to yellow in colour, comparable to hallowe'en pumpkins.

Since the late 1980s, baby carrots or mini-carrots (carrots that have been peeled and cut into uniform cylinders) have been a popular ready-to-eat snack food available in many supermarkets.

The sweetness of carrots allows the vegetable to be used in some fruit-like roles. Grated carrots are used in carrot cakes, as well as carrot puddings, an old English dish thought to have originated in the early 19th century. Carrots can also be used alone or with fruits in jam and preserves. Carrot juice is also widely marketed, especially as a health drink, either stand-alone or blended with fruits and other vegetables.

Companion Plant

Carrots are useful companion plants for gardeners. There is experimental evidence that growing it intercropped with tomatoes increases tomato production. If left to flower, it (like any umbellifer) attracts predatory wasps which kill many garden pests.

History

The wild ancestors of the carrot are likely to have come from Iran and Afghanistan, which remains the centre of diversity of *D. carota*, the wild carrot. Selective breeding over the centuries of a naturally occurring subspecies of the wild carrot, *Daucus carota* subsp. *sativus*, to reduce bitterness, increase sweetness and minimise the woody core, has produced the familiar garden vegetable. In early use, carrots were grown for their aromatic leaves and seeds, not their roots. Some relatives of the carrot are still grown for these, such as parsley, fennel, dill and cumin. The first mention of the root in classical sources is in the 1st century.

The modern carrot appears to have been introduced to Europe in the 8–10th centuries. The 12th century. Arab Andalusian agriculturist, Ibn al-'Awwam, describes both red and yellow carrots; Simeon Seth also mentions both colours in the 11th century. Orange-coloured carrots appeared in the Netherlands in the 17th century. These, the modern carrots, were intended by the antiquary John Aubrey (1626–1697) when he noted in his memoranda "Carrots were first sown at Beckington in Somersetshire Some very old Man there [in 1668] did remember their first bringing hither."

In addition to wild carrot, these alternative (mostly historical) names are recorded for *Daucus carota*: bee's-nest, bee's-nest plant, bird's-nest, bird's-nest plant, bird's-nest root, carota, carotte (French), carrot, common carrot, crow's-nest, daucon, dawke, devil's-plague, fiddle, gallicam, garden carrot, gelbe Rübe (German), gingidium, hill-trot, laceflower, mirrot, Möhre (German), parsnip (misapplied), Queen Anne's lace, rantipole, staphylinos, and zanahoria (Spanish).

Chemistry

Polyacetylenes can be found in Apiaceae vegetables like carrots where they show cytotoxic activities. Falcarinol and falcarindiol (cisheptadeca-1,9-diene-4,6-diyne-3,8-diol) are such compounds. This latter compound shows antifungal activity towards *Mycocentrospora acerina* and *Cladosporium cladosporioides*.

Other compounds such as 6-hydroxymellein, 6-methoxymellein, eugenin, 2,4,5-trimethoxybenzaldehyde (gazarin) or (Z)-3-acetoxy-heptadeca-1,9-diene-4,6-diin-8-ol (falcarindiol 3-acetate) can also be found in carrot. Falcarindiol is the main compound responsible for bitterness in carrots.Cultivars

Carrot cultivars can be grouped into two broad classes, eastern carrots and western carrots. More recently, a number of novelty cultivars have been bred for particular characteristics. The city of Holtville, California, promotes itself as "Carrot Capital of the World", and holds an annual festival devoted entirely to the carrot.

Eastern Carrots

Eastern carrots were domesticated in Central Asia, probably in modern-day Iran and Afghanistan in the 10th century, or possibly earlier. Specimens of the eastern carrot that survive to the present day are commonly purple or yellow, and often have branched roots. The purple colour common in these carrots comes from anthocyanin pigments.

Western Carrots

The western carrot emerged in the Netherlands in the 17th century, from Iran with violet colour, its orange colour making it popular in those countries as an emblem of the House of Orange and the struggle for Dutch independence. The orange colour results from abundant carotenes in these cultivars.

Western carrot cultivars are commonly classified by their root shape:

- Chantenay carrots are shorter than other cultivars, but have greater girth, sometimes growing up to 8 centimetres (3 in) in diametre. They have broad shoulders and taper towards a blunt, rounded tip. They are most commonly diced for use in canned or prepared foods.
- Danvers carrots have a conical shape, having well-defined shoulders and tapering to a point at the tip. They are somewhat shorter than Imperator cultivars, but more tolerant of heavy soil. Danvers cultivars are often puréed as baby food. They were developed in 1871 in Danvers, Ma.

One particular variety lacks the usual orange pigment from carotenes, owing its white colour to a recessive gene for tocopherol (vitamin E). Derived from *Daucus carota* L. and patented at the University of Wisconsin–Madison, the variety is intended to supplement the dietary intake of Vitamin E.

Production Trends

In 2009, according to the Food and Agriculture Organisation, 33.582 million tonnes of carrots and turnips were produced worldwide. With 15.168 million tonnes, China was by far the largest producer and accounted for 45.2 % of the global output, followed by Russia (1.518 million tonnes) and the United States (1.304 million tonnes).

Storage

Carrots can be stored for several months in the refrigerator or over winter in moist sand in a cool place.

2

Classification and Identification of Vegetables

Cauliflower

Cauliflower is one of several vegetables in the species *Brassica oleracea*, in the family Brassicaceae. It is an annual plant that reproduces by seed. Typically, only the head (the *white curd*) is eaten. The cauliflower head is composed of a white inflorescence meristem. Cauliflower heads resemble those in broccoli, which differs in having flower buds. Its name is from Latin *caulis* (cabbage) and flower,. *Brassica oleracea* also includes cabbage, brussels sprouts, kale, broccoli, and collard greens, though they are of different cultivar groups. For such a highly modified plant, cauliflower has a long history. François Pierre La Varenne employed *chouxfleurs* in *Le cuisinier françois*. They had been introduced to France from Genoa in the 16th century, and are featured in Olivier de Serres' *Théâtre de l'agriculture* (1600), as *cauli-fiori* "as the Italians call it, which are still rather rare in France; they hold an honorable place in the garden because of their delicacy", but they did not commonly appear on grand tables until the time of Louis XIV.

Major Groups

There are four major groups of cauliflower.

Italian

Diverse in appearance, and biennial and annual in type, this group includes white, Romanesco, various green, purple, brown and yellow cultivars. This type is the ancestral form from which the others were derived.

Northwest European biennial

Used in Europe for winter and early spring harvest, this was developed in France in the 19th century, and includes the old cultivars Roscoff and Angers.

Northern European annuals

Used in Europe and North America for summer and fall harvest, it was developed in Germany in the 18th century, and includes old cultivars Erfurt and Snowball.

Asian

A tropical cauliflower used in China and India, it was developed in India during the 19th century from the now-abandoned Cornish type, and includes old varieties Early Patna and Early Benaras.

Varieties

There are hundreds of historic and current commercial varieties used around the world. A comprehensive list of about 80 North American varieties is maintained at North Carolina State University.

Colours

White

White cauliflower is the most common colour of cauliflower.

Orange

Orange cauliflower (*B. oleracea* L. var. *botrytis*) contains 25 times the level of vitamin A of white varieties. This trait came from a natural mutant found in a cauliflower field in Canada. Cultivars include 'Cheddar' and 'Orange Bouquet'.

Green

Green cauliflower, of the *B. oleracea* botrytis group, is sometimes called broccoflower. It is available both with the normal curd shape and a variant spiky curd called Romanesco broccoli. Both types have been commercially available in the U.S. and Europe since the early 1990s. Green-curded varieties include 'Alverda', 'Green Goddess' and 'Vorda'. Romanesco varieties include 'Minaret' and 'Veronica'.

Purple

Purple colour in cauliflower is caused by the presence of the antioxidant group anthocyanins, which can also be found in red cabbage and red wine. Varieties include 'Graffiti' and

'Purple Cape'. In Great Britain and southern Italy, a broccoli with tiny flower buds is sold as a vegetable under the name "purple cauliflower". It is not the same as standard cauliflower with a purple curd.

Nutrition

Cauliflower is low in fat, low in carbohydrates but high in dietary fiber, folate, water, and vitamin C, possessing a high nutritional density.

Cauliflower contains several phytochemicals, common in the cabbage family, that may be beneficial to human health.

- Sulforaphane, a compound released when cauliflower is chopped or chewed, may protect against cancer.
- Other glucosinolates
- Carotenoids
- Indole-3-carbinol, a chemical that enhances DNA repair, and acts as an estrogen antagonist, slowing the growth of cancer cells.

Boiling reduces the levels of these compounds, with losses of 20–30% after five minutes, 40–50% after ten minutes, and 75% after thirty minutes. However, other preparation methods, such as steaming, microwaving, and stir frying, have no significant effect on the compounds. A high intake of cauliflower has been associated with reduced risk of aggressive prostate cancer.

Cooking

Cauliflower can be roasted, boiled, fried, steamed or eaten raw. Steaming or microwaving better preserves anticancer compounds than boiling. When cooking, the outer leaves and thick stalks are removed, leaving only the florets. The leaves are also edible, but are most often discarded.

The florets should be broken into similar-sized pieces so they are cooked evenly. After eight minutes of steaming, or five minutes of boiling, the florets should be soft, but not mushy (depending on size). Stirring while cooking can break the florets into smaller, uneven pieces. Cauliflower is often served with a cheese sauce, as in the dish cauliflower cheese. Low carbohydrate dieters can use cauliflower as a reasonable substitute for potatoes; while they can produce a similar texture, or mouth feel, they lack the starch of potatoes.

Fractal Dimension

Cauliflower has been noticed by mathematicians for its distinct fractal dimension, predicted to be about 2.8.

History

The first reliable reference to cauliflower is found in the writings of the Arab Muslim scientists Ibn al-'Awwam and Ibn al-Baitar, between the 12th and 13th centuries.Kohlrabi. Kohlrabi (German turnip) (*Brassica oleracea* Gongylodes group) is a low, stout cultivar of cabbage that will grow almost anywhere. The name comes from the German *Kohl* ("cabbage") plus *Rübe* ~ *Rabi* (Swiss German variant) ("turnip"), because the swollen stem resembles the latter, hence its Austrian name *Kohlrübe*.

Kohlrabi has been created by artificial selection for lateral meristem growth (a swollen, nearly spherical shape); its origin in nature is the same as that of cabbage, broccoli, cauliflower, kale, collard greens, and Brussels sprouts: they are all bred from, and are the same species as the wild cabbage plant (*Brassica oleracea*).

The taste and texture of kohlrabi are similar to those of a broccoli stem or cabbage heart, but milder and sweeter, with a higher ratio of flesh to skin. The young stem in particular can be as crisp and juicy as an apple, although much less sweet.

Figure: *Kohlrabi against an exotic cushion, Gurgaon, India*

Except for the Gigante cultivar, spring-grown kohlrabi much over 5 cm in size tend to be woody, as do full-grown kohlrabi much over perhaps 10 cm in size; the Gigante cultivar can achieve great size while remaining of good eating quality. The plant matures in 55–60 days after sowing. Approximate weight is 150 g and has good standing ability for up to 30 days after maturity.

Kohlrabi can be eaten raw as well as cooked. There are several varieties commonly available, including White Vienna, Purple Vienna, Grand Duke, Gigante (also known as "Superschmelz"), Purple Danube, and White Danube. Colouration of the purple types is superficial: the edible parts are all pale yellow. The leafy greens can also be eaten.

Kohlrabi is one of the most commonly eaten vegetables in Kashmir. Locally called *monj*, the vegetable is eaten along with the leaves (*haakh*). A Kashmiri household may have this on their dinner or lunch plates three to four times a week. *Monj* (kohlrabi) is made in many forms. There is a spicy version which the Pandits call *dum monj*, while as the nonspicy version is called *monj-haakh*. In India, Kohlrabi is more commonly called *Knolkhol* (English) or *Nookal* (Hindi). It is also used extensively in Southern part of India. In Kannada, Kohlrabi is called Gedde Kosu or Navilu Kosu.

Preparation

Kohlrabi stems are surrounded by two distinct fibrous layers that do not soften appreciably when cooked. These layers are generally peeled away prior to cooking or serving raw, with the result that the stems often provide a smaller amount of food than one might assume from their intact appearance. Kohlrabi leaves are edible and can be used interchangeably with collard and kale.

Radish

The radish (*Raphanus sativus*) is an edible root vegetable of the Brassicaceae family that was domesticated in Europe, in pre-Roman times. They are grown and consumed throughout the world. Radishes have numerous varieties, varying in size, colour and duration of required cultivation time. There are some radishes that are grown for their seeds; oilseed radishes are grown, as the name implies, for oil production. Radish can sprout from seed to small plant in as little as 3 days.

History

The descriptive Greek name of the genus *Raphanus* means "quickly appearing" and refers to the rapid germination of these plants. *Raphanistrum*, from the same Greek root, is an old name once used for this genus. The common name "radish" is derived from Latin *radix* (root). Although the radish was a well-established crop in Hellenistic and Roman times, which leads to the assumption that it was brought into cultivation at an earlier time, Zohary and Hopf note that "there are almost no archeological records available "to help determine its earlier history and domestication. Wild forms of the radish and its relatives, the mustards and turnip, can be found over west Asia and Europe, suggesting that their domestication took place somewhere in that area. However Zohary and Hopf conclude, "Suggestions as to the origins of these plants are necessarily based on linguistic considerations."

Cultivation

Radishes grow best in full sun and light, sandy loams with pH 6.5–7.0. They are in season from April to June and from October to January in most parts of North America; in Europe and Japan they are available year-round due to the plurality of varieties grown. Summer radishes mature rapidly, with many varieties germinating in 3–7 days, and reaching maturity in three to four weeks. Harvesting periods can be extended through repeated plantings, spaced a week or two apart.

As with other root crops, tilling the soil to loosen it up and remove rocks helps the roots grow. However, radishes are used in no-till farming to help reverse compaction. Most soil types will work, though sandy loams are particularly good for winter and spring crops, while soils that form a hard crust can impair growth. The depth at which seeds are planted affects the size of the root, from 1 cm (0.4 in) deep recommended for small radishes to 4 cm (1.6 in) for large radishes.

Radishes are a common garden crop in the US, and the fast harvest cycle makes them a popular choice for children's gardens. In temperate climates, it's customary to plant radishes every two weeks from early spring until a few weeks before the first frost, except during periods of hot weather. In warm-weather climates, they are normally planted in the fall.

Companion Plant

Radishes serve as companion plants for many other species, because of their ability to function as a trap crop against pests like flea beetles. These pests will attack the leaves, but the root remains healthy and can be harvested later.

Varieties

Broadly speaking, radishes can be categorized into four main types (summer, fall, winter, and spring) and a variety of shapes lengths, colours, and sizes, such as red, pink, white, gray-black or yellow radishes, with round or elongated roots that can grow longer than a parsnip.

Spring or Summer Radishes

Sometimes referred to as European radishes or spring radishes if they are planted in cooler weather, summer radishes are generally small and have a relatively short 3–4 week cultivation time.

- The April Cross is a giant white radish hybrid that bolts very slowly.
- Bunny Tail is an heirloom variety from Italy, where it is known as 'Rosso Tondo A Piccola Punta Bianca'. It is slightly oblong, mostly red, with a white tip.
- Cherry Belle is a bright red-skinned round variety with a white interior. It is familiar in North American supermarkets.
- Champion is round and red-skinned like the Cherry Belle, but with slightly larger roots, up to about 5 cm (2 in), and a milder flavour.
- Red King has a mild flavour, with good resistance to club root, a problem that can arise from poor drainage.
- Sicily Giant is a large heirloom variety from Sicily. It can reach up to two inches in diameter.
- Snow Belle is an all-white variety of radish, similar in shape to the Cherry Belle.
- White Icicle or just Icicle is a white carrot-shaped variety, around 10–12 cm (4–5 in) long, dating back to the 16th century. It slices easily, and has better than average resistance to pithiness.
- French Breakfast is an elongated red-skinned radish with a white splash at the root end. It is typically slightly milder than other summer varieties, but is among the quickest to turn pithy.
- Plum Purple a purple-fuchsia radish that tends to stay crisp longer than average.
- Gala and Roodbol are two varieties popular in the Netherlands in a breakfast dish, thinly sliced on buttered bread.
- Easter Egg is not an actual variety, but a mix of varieties with different skin colours, typically including white, pink, red, and purple radishes. Sold in markets or seed packets under the name, the seed mixes can extend harvesting duration from a single planting, as different varieties may mature at different times.

Winter Varieties

Black Spanish or Black Spanish Round occur in both round and elongated forms, and are sometimes simply called the black radish or known by the French name *Gros Noir d'Hiver*. It dates in Europe to 1548, and was a common garden variety in England and France during the early 19th century. It has a rough black skin with hot-flavoured white flesh, is round or irregularly pear shaped, and grows to around 10 cm (4 in) in diametre.

Daikon refers to a wide variety of winter radishes from Asia. While the Japanese name daikon has been adopted in English, it is also sometimes called the Japanese radish, Chinese radish, Oriental radish or *mooli* (in India and South Asia). Daikon commonly have elongated white roots, although many varieties of daikon exist. One well known variety is April Cross, with smooth white roots. *The New York Times* describes Masato Red and Masato Green varieties as extremely long, well suited for fall planting and winter storage. The Sakurajima daikon is a hot-flavoured variety which is typically grown to around 10 kg (22 lb), but which can grow to 30 kg (66 lb) when left in the ground.

Seed Pod Varieties

The seeds of radishes grow in siliques (widely referred to as "pods"), following flowering that happens when left to grow past their normal harvesting period. The seeds are edible, and are sometimes used as a crunchy, spicy addition to salads. Some varieties are grown specifically for their seeds or seed pods, rather than their roots. The Rat-tailed radish, an old European variety thought to have come from East Asia centuries ago, has long, thin, curly pods which can exceed 20 cm (8 in) in length. In the 17th century, the pods were often pickled and served with meat. The München Bier variety supplies spicy seed pods that are sometimes served raw as an accompaniment to beer in Germany.

Nutritional Value

Radishes are rich in ascorbic acid, folic acid, and potassium. They are a good source of vitamin B_6, riboflavin, magnesium, copper, and calcium. One cup of sliced red radish bulbs provides approximately 20 cal largely from carbohydrates.

Uses

Cooking: The most commonly eaten portion is the napiform taproot, although the entire plant is edible and the tops can be used as a leaf vegetable. It can also be eaten as a sprout.

The bulb of the radish is usually eaten raw, although tougher specimens can be steamed. The raw flesh has a crisp texture and a pungent, peppery flavour, caused by glucosinolates and the enzyme myrosinase which combine when chewed to form allyl isothiocyanates, also present in mustard, horseradish, and wasabi. Radishes may be used in salads, as well as in many European dishes.

Industry: The seeds of the *Raphanus sativus* species can be pressed to extract seed oil. Wild radish seeds contain up to 48% oil content, and while not suitable for human consumption the oil is a potential source of biofuel. The oilseed radish grows well in cool climates.

Culture

Citizens of Oaxaca, Mexico, celebrate the radish in a festival called Noche de los Rábanos (Night of the Radishes) on December 23 as a part of Christmas celebrations. Locals carve religious and popular figures out of radishes and display them in the town square.

Turnip

The turnip or white turnip (*Brassica rapa* subsp. *rapa*) is a root vegetable commonly grown in temperate climates worldwide for its white, bulbous taproot. Small, tender varieties are grown for human consumption, while larger varieties are grown as feed for livestock.

In the north of England and Scotland, the name turnip, shortened to "neeps", often refers to the larger, yellow rutabaga root vegetable which is also known as the "swede" (from "Swedish turnip").

Description

The most common type is mostly white-skinned apart from the upper 1–6 centimetres, which protrude above the ground and are purple, red, or greenish wherever sunlight has fallen. This above-ground part develops from stem tissue, but is fused with the root.

The interior flesh is entirely white. The entire root is roughly conical, but can be occasionally global, about 5–20 centimetres in diametre, and lacks side roots. The taproot (the normal root below the swollen storage root) is thin and 10 centimetres or more in length; it is trimmed off before marketing. The leaves grow directly from the above-ground shoulder of the root, with little or no visible crown or neck (as found in rutabagas).

Turnip leaves are sometimes eaten as "turnip greens" ("turnip tops" in the UK), and they resemble mustard greens in flavour.

Turnip greens are a common side dish in southeastern US cooking, primarily during late fall and winter. Smaller leaves are preferred; however, any bitter taste of larger leaves can be reduced by pouring off the water from initial boiling and replacing it with fresh water. Varieties specifically grown for the leaves resemble mustard greens more than those grown for the roots, with small or no storage roots. Varieties of *B. rapa* that have been developed only for the use of leaves are called Chinese cabbage. Both leaves and root have a pungent flavour similar to raw cabbage or radishes that becomes mild after cooking.

Turnip roots weigh up to about one kilogram, although they can be harvested when smaller. Size is partly a function of variety and partly a function of the length of time the turnip has grown. Most very small turnips (also called baby turnips) are speciality varieties. These are only available when freshly harvested and do not keep well. Most baby turnips can be eaten whole, including their leaves. Baby turnips come in yellow-, orange-, and red-fleshed varieties as well as white-fleshed. Their flavour is mild, so they can be eaten raw in salads like radishes and other vegetables.

Nutrition

The turnip's root is high only in vitamin C. The green leaves of the turnip top ("turnip greens") are a good source of vitamin A, folate, vitamin C, vitamin K and calcium. Turnip greens are high in lutein (8.5 mg / 100g).

One medium raw turnip (122g) contains the following nutritional information according to the USDA:

- Calories : 34
- Fat: 0.12
- Carbohydrates: 7.84
- Fibers: 2.2
- Protein: 1.10
- Cholesterol: 0

Like rutabaga, turnip contains bitter cyanoglucosides that release small amounts of cyanide. Sensitivity to the bitterness of these cyanoglucosides is controlled by a paired gene. Subjects who have inherited two copies of the "sensitive" gene find turnips twice as bitter as those who have two "insensitive" genes, and thus may find turnips and other cyanoglucoside-containing foods intolerably bitter.

Origin

The turnip was a well-established crop in Hellenistic and Roman times, which leads to the assumption that it was brought into cultivation earlier. Sappho, the 7th century BCE Greek poet, calls one of her paramours *Gongýla*, "turnip". Zohary and Hopf note, however, "there are almost no archaeological records available" to help determine its earlier history and domestication. Wild forms of the hot turnip and its relatives the mustards and radishes are found over west Asia and Europe, suggesting their domestication took place somewhere in that area. However Zohary and Hopf conclude, "Suggestions as to the origins of these plants are necessarily based on linguistic considerations."

Cultivation

The 1881 *Household Cyclopedia* gives these instructions for field cultivation of turnips: The benefits derived from turnip husbandry are of great magnitude; light soils are cultivated with profit and facility; abundance of food is provided for man and beast; the earth is turned to the uses for which it is physically calculated, and by being suitably cleaned with this preparatory crop, a bed is provided for grass seeds, wherein they flourish and prosper with greater vigor than after any other preparation.

The first ploughing is given immediately after harvest, or as soon as the wheat seed is finished, either in length or across the field, as circumstances may seem to require.

In this state the ground remains till the oat seed is finished, when a second ploughing is given to it, usually in a contrary direction to the first. It is then repeatedly harrowed, often rolled between the harrowings and every particle of root-weeds carefully picked off with the hand; a third ploughing is then bestowed, and the other operations are repeated. In this stage, if the ground has not been very foul, the seed process.

The next part of the process is the sowing of the seed; this may be performed by drilling machines of different sizes and constructions, through all acting on the same principle. A machine drawn by a horse in a pair of shafts, sows two drills at a time and answers extremely well, where the ground is flat, and the drills properly made up.

The weight of the machine ensures a regularity of sowing hardly to be gained by those of a different size and construction.

From two to three pounds of seed are sown upon the acre (2 to 3 kg/hectare), though the smallest of these quantities will give many more plants in ordinary seasons than are necessary; but as the seed is not an expensive article the greater part of farmers incline to sow thick, which both provides against the danger of part of the seed perishing, and gives the young plants an advantage at the outset.

Turnips are sown from the beginning to the end of June, but the second and third weeks of the month are, by judicious farmers, accounted the most proper time. Some people have sown as early as May, and with advantage, but these early fields are apt to run to seed before winter, especially if the autumn be favourable to vegetation. As a general rule it may be laid down that the earliest sowings should be on the latest soils; plants on such soils are often long before they make any great progress, and, in the end, may be far behind those in other situations, which were much later sown. The hot turnip plant, indeed, does not thrive rapidly till its roots reach the dung, and the previous nourishment afforded them is often so scanty as to stunt them altogether before they get so far. The first thing to be done in this process is to run a horse-hoe, called a scraper, along the intervals, keeping at such a distance from the young plants that they shall not be injured; this operation destroys all the annual weeds which have sprung up, and leaves the plants standing in regular stripes or rows. The hand hoeing then commences, by which the turnips are all singled out at a distance of from 8–12 inches, and the redundant ones drawn into the spaces between the rows.

The singling out of the young plants is an operation of great importance, for an error committed in this process can hardly be afterward rectified. Boys and girls are always employed as hoers; but a steady and trusty man-servant is usually set over them to see that the work is properly executed.

In eight or ten days, or such a length of time as circumstances may require, a horse-hoe of a different construction from the scraper is used. This, in fact, is generally a small plough, of the same kind with that commonly wrought, but of smaller dimensions. By this implement, the earth is pared away from the sides of the drills, and a sort of new ridge formed in the middle of the former interval.

The hand-hoers are again set to work, and every weed and superfluous turnip is cut up; afterward the horse-hoe is employed to separate the earth, which it formerly threw into the furrows, and lay it back to the sides of the drills. On dry lands this is done by the scraper, but where the least tendency to moisture prevails, the small plough is used, in order that the furrows may be perfectly cleaned out. This latter mode, indeed, is very generally practiced. As a root crop, turnips grow best in cool weather; hot temperatures cause the roots to become woody and bad-tasting. They are typically planted in the spring in cold-weather climates (such as the northern US and Canada) where the growing season is only 3-4 months. In temperate climates (ones with a growing season of 5-6 months), turnips may also be planted in late summer for a second fall crop. In warm-weather climates (7 or more month growing season), they are planted in the fall. 55-60 days is the average time from planting to harvest.

Turnips are a biennial plant, taking two years from germination to reproduction. The root spends the first year growing and storing nutrients, and the second year flowers, produces seeds, and dies. The flowers of the turnip are tall and yellow, with the seeds forming in pea-like pods. In areas with less than 7 month growing seasons, temperatures are too cold for the roots to survive the winter months. In order to produce seeds, it is necessary to pull the turnip and store it over winter, taking care not to damage the leaves. During the spring, it may be set back in the ground to complete its life cycle.

Human Use

Pliny the Elder considered the turnip one of the most important vegetables of his day, rating it "directly after cereals or at all events after the bean, since its utility surpasses that of any other plant." Pliny praises it as a source of fodder for farm animals, and this vegetable is not particular about the type of soil it grows in and because it can be left in the ground until the next harvest, it "prevents the effects of famine" for humans (*N.H.* 18.34).

The Macomber turnip is featured in one of the very few historic markers for a vegetable, on Main Road in Westport, Massachusetts. In Nordic countries, turnip was the staple crop before its replacement with potato in the 18th century. The cross between turnip and cabbage, rutabaga, was possibly first produced in Scandinavia.

In England, around 1700, Turnip Townshend promoted the use of turnips in a four-year crop rotation system that enabled year-round livestock production. In the Southern United States, stewed turnips are eaten as a root vegetable in the autumn and winter.

The leaves or "greens" of the turnip are harvested and eaten all year. Turnip Greens are cooked with a ham hock or piece of fat pork meat, the juice produced in the stewing process is prized as "pot liquor". Stewed turnip greens are often eaten with vinegar. In the Tyrolean Alps of Austria, raw shredded Turnip root is served in a chilled remoulade in the absence of other fresh greens as a winter salad.

In the south of England and Scotland, the smaller white vegetable is often called *neeps* or *turnips*, while it is the larger rutabagas which are referred to as *swedes*. Turnips or neeps are mashed and eaten with haggis, traditionally on Burns Night. Turnip lanterns are an old tradition; since inaugural Halloween festivals in Ireland and Scotland, turnips (rutabaga) have been carved out and used as candle lanterns.

At Samhain, candle lanterns carved from turnips — *samhnag* — were part of the traditional Celtic festival. Large turnips were hollowed out, carved with faces and placed in windows, used to ward off harmful spirits. At Halloween in Scotland in 1895, masqueraders in disguise carried lanterns made out of scooped out turnips. Laurie Lee, in "Cider with Rosie", an autobiography of a childhood in the Cotswolds, mentions the Parochial Church Tea and Annual Entertainment, which took place around Twelfth night. "We...saw his red face lit like a hot turnip lamp as he crouched to stoke up the flames."

In Iran, boiled turnip roots (with salt) are a common household remedy for fever. In the Punjab and Kashmir regions (India, Pakistan), turnips are used in variety of dishes. The most famous of these dishes is *shab-daig*.

Heraldry

The turnip is an old vegetable charge in heraldry. It was used by Leonhard von Keutschach, prince-archbishop of Salzburg. The turnip is still the heart shield in the arms of Keutschach am See.

Winter Melon

The winter melon, also called white gourd, winter gourd, ash gourd, "fuzzy gourd", or "fuzzy melon", is a vine grown for its very large fruit, eaten as a vegetable when mature. It is the only member of the genus *Benincasa*. The fruit is fuzzy when young. The immature melon has thick white flesh that is sweet when eaten. By maturity, the fruit loses its hairs and develops a waxy coating, giving rise to the name wax gourd, and providing a long shelf life. The melon may grow as large as 80 cm in length. Although the fruit is referred to as a "melon", the fully grown crop is not sweet. Originally cultivated in Southeast Asia, the winter melon is now widely grown in East Asia and South Asia as well. Winter melon is also a common name for members of the Inodorus cultivar group of the muskmelon (*Cucumis melo* L), more commonly known as casaba or honeydew melons.

Uses

The winter melon requires very warm weather to grow but can be stored for many months much like winter squash. It is commonly eaten throughout winter in countries of deciduous vegetation such as China, as one of the few vegetables available during winter, hence its Chinese name literally means 'winter melon'. The winter melon can typically be stored for 12 months. In Vietnamese cuisine, it is called *bí ðao*, which is usually used to make soup or stew. When cooked with port short ribs, the soup it can help produce more milk for breastfeeding mothers.

In Chinese cuisine the melons are used in stir fry or usually combined with pork or pork/beef bones to make winter melon soup, often served in the scooped out melon, carved by scraping off the waxy coating. It is also chopped and candied as wintermelon candy (táng dông guâ) to be commonly eaten at New Year festivals, or as filling for Sweetheart cake (lÎopó bÐng). It has also been used as the base filling in Chinese and Taiwanese mooncakes for the Moon Festival.

Winter melon is called kundol, kondol, or gondol in the Philippines. It is candied (referred to plainly as "kundol") and is used as a pastry filing for hopia. It is also an ingredient in some savory soups (sabaw) and stir-fries (guisado). It is one of the vegetables mentioned in the Filipino folk song "Bahay Kubo". In North India and Pakistan, the vegetable is also used to prepare a candy called *Petha*. In South Indian cuisine, it is used to make curries. In Ayurvedic remedies it used to increase appetite also its fresh juice is used to cure kidney stones. The seeds are cooked in milk and taken to increase "sperm count" and to improve sperm locomotion.

Occasionally, it is used to produce a fruit drink which has a very distinctive taste. It is usually sweetened with caramelized sugar, which enhances the taste. In Southeast Asia, the drink is widely marketed as winter melon tea. The shoots, tendrils, and leaves of the plant may also be eaten as greens. In India, Ash gourd is used to make a liquefied dish with curds or buttermilk (a popular traditional South Indian recipe).

Bitter Melon

Momordica charantia, called bitter melon, bitter gourd or bitter squash in English, Karavella in Sanskrit and Karela in Hindi and Urdu, Karla in Bengali and Marathi, Pavakai in Tamil, Hagala kayi in Kannada, Kakarakaya in Telugu, kudhreth narhy (kudret narý) in

Turkish, is a tropical and subtropical vine of the family Cucurbitaceae, widely grown in Asia, Africa, and the Caribbean for its edible fruit, which is among the most bitter of all fruits. Its many varieties differ substantially in the shape and bitterness of the fruit. This is a plant of the tropics. Karela originated in India and it was carried to China in the 14th century.

Description

Figure: *Ripe fruit*

This herbaceous, tendril-bearing vine grows to 5 m. It bears simple, alternate leaves 4–12 cm across, with three to seven deeply separated lobes. Each plant bears separate yellow male and female flowers. In the northern hemisphere, flowering occurs during June to July and fruiting during September to November.

The fruit has a distinct warty exterior and an oblong shape. It is hollow in cross-section, with a relatively thin layer of flesh surrounding a central seed cavity filled with large, flat seeds and pith. The fruit is most often eaten green, or as it is beginning to turn yellow. At this stage, the fruit's flesh is crunchy and watery in texture, similar to cucumber, chayote or green bell pepper, but bitter. The skin is tender and edible. Seeds and pith appear white in unripe fruits; they are not intensely bitter and can be removed before cooking. As the fruit ripens, the flesh (rind) becomes tougher, more bitter, and too distasteful to eat.

On the other hand, the pith becomes sweet and intensely red; it can be eaten uncooked in this state, and is a popular ingredient in some Southeast Asian salads. When the fruit is fully ripe, it turns orange and mushy, and splits into segments which curl back dramatically to expose seeds covered in bright red pulp.

Varieties

Bitter melon comes in a variety of shapes and sizes. The China phenotype is 20–30 cm long, oblong with bluntly tapering ends and pale green in colour, with a gently undulating, warty surface. The bitter melon more typical of India has a narrower shape with pointed ends, and a surface covered with jagged, triangular "teeth" and ridges. It is green to white in colour. Between these two extremes are any number of intermediate forms. Some bear miniature fruit of only 6–10 cm in length, which may be served individually as stuffed vegetables. These miniature fruit are popular in India and elsewhere in Southeast Asia.

Culinary Uses

Bitter melon is generally consumed cooked in the green or early yellowing stage. The young shoots and leaves of the bitter melon may also be eaten as greens. Bitter melon is often used in Chinese cooking for its bitter flavour, typically in stir-fries (often with pork and *douchi*), soups, and also as tea. It has also been used in place of hops as the bittering ingredient in some Chinese beers.

It is very popular throughout South Asia. In northern India, it is often prepared with potatoes and served with yogurt on the side to offset the bitterness, or used in *sabzi*. In North Indian cuisine, it is stuffed with spices and then cooked in oil. In southern India, it is used in the dishes *thoran/thuvaran* (mixed with grated coconut), *theeyal* (cooked with roasted coconut) and *pachadi* (which is considered a medicinal food for diabetics).

Other popular recipes include preparations with curry, deep fried with peanuts or other ground nuts, and *Pachi Pulusu*, a soup with fried onions and other spices. In Tamil Nadu, a special preparation in Brahmins' cuisine called *pagarkai pitla* a kind of sour *koottu* variety is very popular.

Also popular is *kattu pagarkkai* a curry stuffed with onions, cooked lentil and grated coconut mix, tied with thread and fried in oil. In

Konkan region of Maharashtra ,salt is added to finely chopped bitter gourd and then it is squeezed, removing its bitter juice to some extent. After frying this with different spices, less bitter and crispy preparation is served with grated coconut. In Pakistan and Bangladesh, bitter melon is often cooked with onions, red chili powder, turmeric powder, salt, coriander powder, and a pinch of cumin seeds. Another dish in Pakistan calls for whole, unpeeled bitter melon to be boiled and then stuffed with cooked ground beef, served with either hot *tandoori* bread, *naan*, *chappati*, or with *khichri* (a mixture of lentils and rice).

Bitter melon is a significant ingredient in Okinawan cuisine, and is increasingly used in mainland Japan. It is popularly credited with Okinawan life expectancies being higher than the already long Japanese ones. In Indonesia, bitter melon is prepared in various dishes, such as *gado-gado*, and also stir fried, cooked in coconut milk, or steamed.

In Vietnam, raw bitter melon slices consumed with dried meat floss and bitter melon soup with shrimp are popular dishes. Bitter melons stuffed with ground pork are served as a popular summer soup in the south. It is also used as the main ingredient of "stewed bitter melon".

This dish is usually cooked for the T¿t holiday, where its "bitter" name is taken as a reminder of the poor living conditions experienced in the past. In the Philippines, bitter melon may be stir-fried with ground beef and oyster sauce, or with eggs and diced tomato.

The dish *pinakbet*, popular in the Ilocos region of Luzon, consists mainly of bitter melons, eggplant, okra, string beans, tomatoes, lima beans, and other various regional vegetables altogether stewed with a little *bagoong*-based stock.

In Nepal, bitter melon is prepared as a fresh pickle called *achar*. For this, the bitter gourd is cut into cubes or slices and sautéed covered in oil and a sprinkle of water. When it is softened and reduced, it is minced in a mortar with a few cloves of garlic, salt and a red or green pepper. It is also sautéed to golden-brown, stuffed, or as a curry on its own or with potatoes. In Trinidad and Tobago, bitter melons are usually sauteed with onion, garlic and scotch bonnet pepper until almost crisp.

Medicinal Uses

Bitter melon has been used in various Asian and African herbal medicine systems for a long time. In Turkey, it has been used as a folk remedy for a variety of ailments, particularly stomach complaints. The fruit is broken up and soaked in either olive oil or honey.

Active Substances

The plant contains several biologically active compounds, chiefly momordicin I and momordicin II, and cucurbitacin B. The plants contains also several bioactive glycosides (including momordin, charantin, charantosides, goyaglycosides, momordicosides) and other terpenoid compounds (including momordicin-28, momordicinin, momordicilin, momordenol, and momordol). It also contains cytotoxic (ribosome-inactivating) proteins such as momorcharin and momordin.

Antihelmintic

Bitter melon is used as a folk medicine in Togo to treat gastrointestinal diseases, and extracts have shown activity *in vitro* against the nematode worm *Caenorhabditis elegans.*

Antimalarial

Bitter melon is traditionally regarded in Asia as useful for preventing and treating malaria. Tea from its leaves is used for this purpose also in Panama and Colombia. In Guyana, bitter melons are boiled and stir-fried with garlic and onions. This popular side dish known as *corilla* is served to prevent malaria. Laboratory studies have confirmed that species related to bitter melon have antimalarial activity, though human studies have not yet been published.

Antiviral

In Togo, the plant is traditionally used against viral diseases such as chickenpox and measles. Tests with leaf extracts have shown *in vitro* activity against the herpes simplex type 1 virus, apparently due to unidentified compounds other than the momordicins.

Laboratory tests suggest compounds in bitter melon might be effective for treating HIV infection. As most compounds isolated from bitter melon that impact HIV have either been proteins or lectins, neither of which are well-absorbed, it is unlikely that oral intake of bitter melon will slow HIV in infected people. Oral ingestion of bitter melon possibly could offset negative effects of anti-HIV drugs, if an *in vitro* study can be shown to be applicable to people.

Cardioprotective

Studies in mice indicate bitter melon seed may have a cardioprotective effect by down-regulating the NF-êB inflammatory pathway.

Diabetes

In 1962, Lolitkar and Rao extracted from the plant a substance, which they called charantin, which had hypoglycaemic effect on normal and diabetic rabbits. Another principle, active only on diabetic rabbits, was isolated by Visarata and Ungsurungsie in 1981. Bitter melon has been found to increase insulin sensitivity. In 2007, a study by the Philippine Department of Health determined a daily dose of 100 mg per kilogram of body weight is comparable to 2.5 mg/kg of the antidiabetes drug glibenclamide taken twice per day. Tablets of bitter melon extract are sold in the Philippines as a food supplement and exported to many countries. Other compounds in bitter melon have been found to activate the AMPK, the protein that regulates glucose uptake (a process which is impaired in diabetics).

Bitter melon also contains a lectin that has insulin-like activity due to its nonprotein-specific linking together to insulin receptors. This lectin lowers blood glucose concentrations by acting on peripheral tissues and, similar to insulin's effects in the brain, suppressing appetite. This lectin is likely a major contributor to the hypoglycemic effect that develops after eating bitter melon.

Anticancer

Two compounds extracted from bitter melon, á-eleostearic acid (from seeds) and 15,16-dihydroxy-á-eleostearic acid (from the fruit) have been found to induce apoptosis of leukemia cells *in vitro*. Diets containing 0.01% bitter melon oil (0.006% as á-eleostearic acid) were found to prevent azoxymethane-induced colon carcinogenesis in rats. Researchers at Saint Louis University claim an extract from bitter melon, commonly eaten and known as *karela* in India, causes a chain of events which helps to kill breast cancer cells and prevents them from multiplying.

Other Uses

Bitter melon has been used in traditional medicine for several other ailments, including dysentery, colic, fevers, burns, painful menstruation, scabies and other skin problems. It has also been used as abortificant, for birth control, and to help childbirth.

Cautions

The seeds of bitter melon contains vicine, so can trigger symptoms of favism in susceptible individuals. In addition, the red arils of the seeds are reported to be toxic to children, and the fruit is contraindicated during pregnancy.

Calabash

Lagenaria siceraria (synonym *Lagenaria vulgaris* Ser.), bottle gourd, opo squash or long melon is a vine grown for its fruit, which can either be harvested young and used as a vegetable, or harvested mature, dried, and used as a bottle, utensil, or pipe. For this reason, the calabash is widely known as the bottle gourd. The fresh fruit has a light green smooth skin and a white flesh. Rounder varieties are called calabash gourds. They come in a variety of shapes, they can be huge and rounded, or small and bottle shaped, or slim and more than a metre long. The calabash was one of the first cultivated plants in the world, grown not primarily for food, but for use as a water container. The bottle gourd may have been carried from Africa to Asia, Europe and the Americas in the course of human migration. It shares its common name with that of the calabash tree (*Crescentia cujete*).

Etymology

The word comes from the Spanish *calabaza*, possibly from Arabic *qar'a yabisa* "dry gourd", from Persian *kharabuz*, used of various large melons; or from a pre-Roman Iberian *calapaccia*.

Origin and Dispersal

It is a commonly cultivated plant in tropical and subtropical areas of the world, now believed by some to have spread or originated from wild populations in southern Africa. Stands of *Lagenaria siceraria* that may be source plants, and not merely domesticated stands run wild, were reported recently in Zimbabwe. This apparent domestication source plant produces thinner-walled fruit that, when dried, would not endure the rigors of use on long journeys as a water container. Today's calabash may owe its tough, waterproof wall to selection pressures over its long history of domestication.

Cultivation

Calabash had been cultivated in Asia, Europe and the Americas for thousands of years before Columbus's discovery of America. Historically, in Europe, Walahfrid Strabo (808–849), abbot and poet from Reichenau, advisor to the Carolingian kings, discussed it in his Latin Hortulus as one of the 23 plants of an ideal garden. Recent research indicates some can have an African origin and at least two unrelated domestications: one 8–9 thousand years ago, based on the analysis of archeological samples found in Asia, a second, four thousand years ago, traced from archeological discoveries in Egypt.

The mystery of the calabash – namely that this African or Eurasian species was being grown in America over 8,000 years ago – came about from the difficulty in understanding how it came to be on the American continent. Genetic research on archeological samples published by the

National Academy of Sciences in December 2005 suggests calabash may have been domesticated earlier than food crops and livestock, and, like dogs, were brought into the New World at the end of the ice age by Paleo-Indians.

It is supposed that bottle gourds were carried by people in boats or on foot across the then-existing land bridge between Asia and America. Once in Florida and Mexico, bottle gourd seeds could still be viable after long periods of migration.

The rind of the domesticated calabash, unlike that of its wild counterpart, is thick and waterproof. Calabash previously was thought to have spread across oceans without human intervention, if the seeds were still able to germinate even after long periods at sea.

This was the basis of the earlier, dominating theory, which proposed the calabash had drifted across the Atlantic Ocean from Africa to North and South America. The new research notes domestication had led to changes in morphology (shape) of Asian and African specimens, potentially allowing the identification of the calabash from different areas. Now, both genetic and morphological considerations show calabash found in American archaeological finds are closer to Asian calabash variants than to African ones.

Toxicity

Like other members of the Cucurbitaceae family, calabashes contain cucurbitacins that are known to be cytotoxic. The tetracyclic triterpenoid cucurbitacins present in fruits and vegetables of the cucumber family, are responsible for the bitter taste, and can cause ulcers in the stomach. In extreme cases, people have died from drinking calabash juice.

Culinary Uses

In India, it is known as *lauki, dudhi* or *ghiya* in Hindi/Urdu/ Gujarati; *Laau* in Oriya; *aal* in Marwari; *churakka* in Malayalam; *jatilao* in Assamese; *lau* in Bengali; *sorakaaya or anapakaya* in Telugu; *dudhi-Bhopala* in Marathi; *sorekayi* in Kannada; *sajmain* in Maithili and *suraikkaai* in Tamil. A popular North Indian dish is *lauki channa,* (channa dal and diced gourd in a semidry gravy). In Maharashtra, the skin of the gourd is used in making a Chutney preparation.

In parts of India, the dried, unpunctured gourd is used as a float (called *surai-kuduvai* in Tamil) to learn swimming in rural areas. Indian musical instruments, such as the tanpura, sitar and rudra veena, are constructed from dried calabash gourds, using special cultivars that were originally imported from Africa and Madagascar. They are mostly grown in Bengal and near Miraj, Maharashtra. These gourds are valuable items and they are carefully tended, sometimes they are given injections to stop worms and insects from making holes while they are drying., etc.

The calabash, as a vegetable, is frequently used in southern Chinese cuisine as either a stir-fry or in a soup. The Chinese name for calabash is *hulu* (simplified Chinese: k„¦,; traditional Chinese: k„†; pinyin: *húlu*) or *huzi* (Chinese: k„P[; pinyin: *húzi*) in Mandarin. Two common kinds of calabash sold in Chinese stores are the "Opo" kind, which is elongated but still plump, and "Mao Gua" which is very similar to Opo, but it has hairs, as its Chinese name references, which translates to "Hairy Squash". The hairs, although small, can get embedded in the skin, but it is usually safe for adults to handle.

In Japan, the species is known as *hyôtan* or *yûgao*, with the former word referring particularly to the larger-fruiting variety whose fruits are used mostly for making containers or other handicrafts and the latter referring to the smaller-fruiting variety whose fruits are more edible.

Names used to refer particularly to the fruit of one or another variety of this species include *fukube* and *hisago*.

It is most commonly sold in the form of dried, marinated strips known as *kanpyô*, and is commonly used as an ingredient for making makizushi (rolled sushi).

In Korea, it is known as *bak* (¼) or *jorongbak*. In Burma, it is known as 08.8 *boo thee*, a popular fruit; young leaves are also boiled and eaten with spicy hot, fermented fish sauce called *nga peet*.

In the Philippines, it is known as *upo*. In Italian cuisine, it is known as *cucuzza* (plural *cucuzze*). In Central America, the seeds of the calabash gourd are toasted and ground with other ingredients (including rice, cinnamon, and allspice) to make the drink horchata. Calabash is known locally as *morro* or *jícaro*. In Colombia and Venezuela, the calabash is known as a *tapara* or *totuma*.

In Pakistan, it is known as *lauki/kaddu* in Urdu. In Bangladesh, it is called *lau*. In Nepali, it is called *lauka*. In Arabic, it is called *qara*. The tender young gourd is cooked as a summer squash.

In Vietnam, it is called *b§u canh* or *b§u n-m*, and is used in a variety of dishes: boiled, stir-fried, soup dishes and as a medicine.

The shoots, tendrils, and leaves of the plant may also be eaten as greens.

Cultural Uses

The Caribbean: Calabash is primarily used for utensils, such as cups, bowls, and basins in rural areas. It can be used for carrying water, or can be made for carrying items, such as fish, when fishing. In some Caribbean countries, it is worked, painted and decorated as shoulder bags or other items by artisans, and sold to tourists. In Jamaica, it is also a reference to the natural lifestyle of Rastafarians.

As a cup, bowl, or even water-pipe or "bong", the calabash is considered consistent with the "Ital" or vital lifestyle of not using refined products such as table salt, or using modern cooking methods, such as microwaves. In Haiti, the plant is called *kalbas kouran,* literally "running calabash", and is used to make the sacred rattle emblematic of the Vodou priesthood, called an *asson.* As such, the plant is highly respected. It is also the national tree of St. Lucia.

Africa

Hollowed out and dried calabashes are a very typical utensil in households across West Africa. They are used to clean rice, carry water and as food containers. Smaller sizes are used as bowls to drink palm wine. Calabashes are used in making the West African kora (a harp-lute), *xalam/ngoni* (a lute) and the *goje* (a traditional fiddle). They also serve as resonators underneath the *balafon* (West African marimba).

The calabash is also used in making the *shegureh* (a Sierra Leonean women's rattle) and *balangi* (a Sierra Leonean type of *balafon*) musical instruments. Sometimes, large calabashes are simply hollowed, dried and used as percussion instruments, especially by Fulani, Songhai, Gur-speaking and Hausa peoples. In Nigeria, the calabash has been used to avoid a law requiring the wearing of a helmet on a motorcycle. In South Africa, it is commonly used as a drinking vessel by tribes such as the Zulus. Ebore tribe children in Ethiopia wear hats made from the calabash to protect them from the sun. Recently, the Soccer City stadium which hosted the FIFA World Cup has been completed and its shape takes inspiration from the calabash.

Mexico: In many rural parts of Mexico, the calabash is dried and carved hollow to create a *bule* or a *guaje,* a gourd used to carry water around like a canteen. The gourd cut in half, called *jícara,* gave the parallel name to a clay cup *jícara.*

Costa Rica: The Costa Rican town of Santa Bárbara de Santa Cruz holds a traditional annual dance of the calabashes (*baile de los guacales*). Since 2000, the activity has been considered of cultural interest to the community, and all participants receive a hand-painted calabash vessel to thank them for their economic contribution (which they paid in the form of an entrance ticket). Aboriginals throughout the country traditionally serve *chicha* in calabash vessels to the participants of special events such as the *baile de los diablitos* (dance of the little fiends).

South America: In Peru, Brazil, Argentina, Uruguay and Paraguay, calabash gourds are dried and carved into *mates*, the traditional container for the popular caffeinated tea-like drink brewed from the yerba mate plant. In Brazil, this container is called *cuia*, *porongo* or *cabaça*. Gourds also commonly used as the resonator for the *berimbau*, the signature instrument of *capoeira*, a martial art/ dance developed in Brazilian plantations by African slaves. The calabash gourd is possibly mankind's oldest instrument resonator.

In Peru, Bolivia and Ecuador, calabash gourds are known to have been used for medicinal purposes for over a thousand years by Andean cultures. The Inca culture applied folklore symbology to gourds to pass down from one generation to another, and this practice is still familiar and valued.

Bowls made of calabash were used by indigenous Brazilians as utensils made to serve food, and the practice is still retained in some remote areas of Brazil (originally by populations of various ethnicities, origins and regions, but nowadays mainly the indigenes themselves).

Venezuela: The president of Venezuela, Hugo Chávez, has suggested Venezuelans avoid showers longer than three minutes. Critics of Chavez have ridiculed this (*reductio ad incommodum*) by ironically suggesting the use of a *totuma* to bathe (although Chavez himself did not suggest this). The inference is that Chavez's bathing suggestion is an unwelcome intrusion into Venezuelans' daily lives, and further, that bathing with a gourd is shamefully primitive. Compare U.S. President Jimmy Carter's speech urging Americans to conserve energy during the US 1979 energy crisis and negative reaction by his critics.

China: The *hulu* is an ancient symbol for health. In the old days, the doctors would carry medicine inside it, so it has fabled properties for healing. The *hulu* is believed to absorb negative earth-based *qi* (energy) that would otherwise affect health, and is a traditional Chinese medicine cure.

Dried calabash is also used as containers of liquids, often liquors or medicine. Calabash gourds were also grown in earthen molds to form different shapes with imprinted floral or arabesque design, and dried to house pet crickets, which were kept for their song and fighting abilities. The texture of the gourd lends itself nicely to the sound of the animal, much like a musical instrument. It is a symbol of the Xian immortals.

India: Hindu ascetics (*sadhu*) traditionally use a dried gourd vessel called the *kamandalu*. The juice of lauki is considered to have many medicinal properties and to be very good for health. The Baul singers of Bengal have their musical instruments made out of it. The practice is also common among Buddhist and Jain sages.

Hawaii

In Hawaii, a calabash is a large serving bowl, usually made from a hardwood rather than from the calabash gourd as in Maroon cultures. It is used on a buffet table or in the middle of the dining table. The use of the calabash in Hawaii has led to terms like "calabash family" or "calabash cousins", indicating an extended family grown up around shared meals and close friendships. Food is very important in modern Hawaiian culture. The expression "*e komo mai* - Welcome/Come in" was the standard welcome to anyone approaching a home. This gourd is often dried when ripe and used as a percussion instrument in contemporary and ancient hula.

Other Uses: Additionally, the gourd can be dried and used to smoke pipe tobacco. A typical design yielded by this squash is recognised (theatrically) as the pipe of Sherlock Holmes, but Doyle never mentioned Holmes using a calabash pipe. It was the preferred pipe for stage actors portraying Holmes, because they could balance this pipe better than other styles while delivering their lines. See, Smoking pipe (tobacco)#Calabash.

Coccinia

The Scarlet gourds (*Coccinia*) is a genus with 28-30 species.. It is distributed in sub-Saharan Africa and with one species in South and Southeast Asia that is also introduced into the New World. It is best known for the cultivated crop *C. grandis*, the Ivy gourd.

Coccinia Grandis

Coccinia grandis, the ivy gourd, also known as *baby watermelon, little gourd* or *gentleman's toes* is a tropical vine. It is also known as *Cephalandra indica* and *Coccinia indica.*

Geographic Spread

Coccinia grandis' native range extends from Africa to Asia, including India, the Philippines, China, Indonesia, Malaysia, Thailand, Vietnam, eastern Papua New Guinea, and the Northern Territories, Australia. Its documented introduced range includes the Federated States of Micronesia, Fiji, Guam, Saipan, Hawaii, the Marshall Islands, Samoa, Tonga, and Vanuatu.

Seeds or fragments of the vine can be relocated and lead to viable offspring. This can occur when humans transport organic debris or equipment containing *C. grandis*. Once the ivy gourd is established, it is presumably spread by birds, rats, and other mammals. In Hawaii, it has been suggested that the fruit may be dispersed by pigs. Long-distance dispersal is most commonly carried out by humans due to its culinary uses or by mistake.

Regarded as very invasive and on the Hawaii State Noxious Weed List, ivy gourd can grow up to four inches per day. It grows in dense blankets, shading other plants from sunlight and high-jacking nutrients, effectively killing vegetation underneath.

Native to tropical Africa and Asia, it was introduced to Hawaii as a backyard food crop. It is sometimes tolerated along garden fences and other outdoor features because of its attractive white flowers. It has escaped to become a vigorous pest in Hawaii, Florida, Australia, and Texas. In Hawaii, this plant has spread quickly through Manoa Valley to Punchbowl and into Waimânalo, Oahu, and into the Kona area.

Weed control

There are both physical and chemical recommendations for control of the ivy gourd. It is very difficult to control this plant physically except by bagging fruits. Hand-harvesting normally does not kill the plant but rather breaks the vine blankets into smaller pieces and the plant is able to reestablish when it touches the ground. These methods can make the infestation worse and further the need for more rigorous control methods. Picking the fruit and placing them in plastic bags can help decrease the seed back that is present with the soil. When utilizing chemical controls, that ivy gourd responded well to a thin-lined bark application of 100% Garlon 4 (triclopyr), leaving plants in place so as not to translocate the herbicide or spread the pest.

It is applied multiple times until the vine dies. In Hawaii several species of insect have been introduced with the purpose of being a biocontrol. Two weevils, *Acythopeus burkhartorum* and *A. cocciniae*, were introduced by the Department of Agriculture to Oahu and Hawaii. African vine moths (*Melittia oedipus*) were also released onto Oahu and Maui. On the island of Maui it appears that the *A. cocciniae* is established and are damaging leaves. The larva feed on the plant and the adults chew holes in the leaves. The moth has yet to appear successful in its purpose.

Cultivation

In Southeast Asia, ivy gourd is grown for its edible young shoots and edible fruits.

Medicinal Value

In traditional medicine, fruits have been used to treat leprosy, fever, asthma, bronchitis and jaundice. The fruit possesses mast cell stabilizing, anti-anaphylactic and antihistaminic potential. Ivy gourd extracts and other forms of the plant can be purchased online and in health food stores. It is claimed that these products help regulate blood sugar levels. There is some research to support that compounds in the plant inhibit the enzyme glucose-6-phosphatase. Glucose-6-phosphatase is one of the key liver enzymes involved in regulating sugar metabolism. Therefore, ivy gourd is sometimes recommended for diabetic patients. Although these claims have not been supported, there currently is a fair amount of research focused on the medicinal properties of this plant focusing on its use as an antioxidant, anti-hypoglycemic agent, immune system modulator, etc. Some countries in Asia like Thailand prepare traditional tonic like drinks for medicinal purposes.

Recipes

There are a variety of recipes from all over the world that list ivy gourd as the main ingredient. It is often compared to bitter melon. The fruit is commonly eaten in Indian cuisine.

People of Indonesia and other Southeast Asian countries also consume the fruit and leaves. Cultivation of ivy gourd in home gardens has been encouraged in Thailand due to its being a good source of several micronutrients, including vitamins A and C.

In India it is eaten as a curry, by deep-frying it along with or without chilli and garlic; stuffing it with masala and sauteing it, or boiling it first in a cooker and then frying it. It is also used in sambar, a vegetable and lentil-based soup. Rashmati are best when cooked. They do not generally taste good when raw.

Cucumber

The cucumber (*Cucumis sativus*) is a widely cultivated plant in the gourd family Cucurbitaceae, which includes squash, and in the same genus as the muskmelon. The plant is a creeping vine which bears cylindrical edible fruit when ripe. There are three main varieties of cucumber: "slicing", "pickling", and "burpless". Within these varieties, several different cultivars have emerged. The cucumber is originally from India but is now grown on most continents. Many different varieties are traded on the global market.

Description

The cucumber is a creeping vine that roots in the ground and grows up trellises or other supporting frames, wrapping around supports with thin, spiraling tendrils. The plant has large leaves that form a canopy over the fruit.

The fruit of the cucumber is roughly cylindrical, elongated with tapered ends, and may be as large as 60 centimetres (24 in) long and 10 centimetres (3.9 in) in diametre. Having an enclosed seed and developing from a flower, botanically speaking, cucumbers are classified as fruits. However, much like tomatoes and squash they are often perceived, prepared and eaten as vegetables. Cucumbers are usually more than 90% water.

Flowering and Pollination

A few varieties of cucumber are parthenocarpic, the blossoms creating seedless fruit without pollination. Pollination for these varieties degrades the quality. In the United States, these are usually grown in greenhouses, where bees are excluded. In Europe, they are grown outdoors in some regions, and bees are excluded from these areas. Most cucumber varieties, however, are seeded and require pollination. Thousands of hives of honey bees are annually carried to cucumber fields just before bloom for this purpose. Cucumbers may also be pollinated by bumblebees and several other bee species.

Symptoms of inadequate pollination include fruit abortion and misshapen fruit. Partially pollinated flowers may develop fruit which are green and develop normally near the stem end, but pale yellow and withered at the blossom end. Traditional varieties produce male blossoms first, then female, in about equivalent numbers. New gynoecious hybrid cultivars produce almost all female blossoms. However, since these varieties do not provide pollen, they must have interplanted a pollenizer variety and the number of beehives per unit area is increased. Insecticide applications for insect pests must be done very carefully to avoid killing off the insect pollinators.

Genome

In 2009, an international team of researchers announced they had sequenced the cucumber genome.

Distribution and Habitat

According to the Food and Agriculture Organisation of the United Nations, China produced at least 60% of the global output of cucumber and gherkin in 2005, followed at a distance by Turkey, Russia, Iran and the United States.

Cultivation

History: Cucumbers originated in India from Cocumis hystrix. where a great many varieties of cucumber have been observed. It has been cultivated for at least 3,000 years, and was probably introduced to other parts of Europe by the Greeks or Romans. Records of cucumber cultivation appear in France in the 9th century, England in the 14th century, and in North America by the mid-16th century.

Earliest Cultivation: The cucumber is listed among the foods of ancient Ur, and the legend of Gilgamesh describes people eating cucumbers. Some sources also state it was produced in ancient Thrace, and it is certainly part of modern cuisine in Bulgaria and Turkey, parts of which make up that ancient state. From India, it spread to Greece (where it was called *síkyon*) and Italy (where the Romans were especially fond of the crop), and later into China.

According to Pliny the Elder (*The Natural History*, Book XIX, Chapter 23), the Ancient Greeks grew cucumbers, and there were different varieties in Italy, Africa, and modern-day Serbia.

Roman Empire: According to Pliny, the Emperor Tiberius had the cucumber on his table daily during summer and winter. The Romans reportedly used artificial methods (similar to the greenhouse system) of growing to have it available for his table every day of the year.

"Indeed, he was never without it; for he had raised beds made in frames upon wheels, by means of which the cucumbers were moved and exposed to the full heat of the sun; while, in winter, they were withdrawn, and placed under the protection of frames glazed with mirrorstone. Reportedly, they were also cultivated in cucumber houses glazed with oiled cloth known as 'specularia'.

Pliny the Elder describes the Italian fruit as very small, probably like a gherkin, describing it as a wild cucumber considerably smaller than the cultivated one. Pliny also describes the preparation of a medication known as *elaterium*, though some scholars believe he was referring to *Cucumis silvestris asininus*, a species different from the common cucumber.

Pliny also writes about several other varieties of cucumber, including the cultivated cucumber, and remedies from the different types (9 from the cultivated, 5 from the "anguine", and 26 from the "wild").

The Romans are reported to have used cucumbers to treat scorpion bites, bad eyesight, and to scare away mice. Wives wishing for children wore them around their waists. They were also carried by the midwives, and thrown away when the child was born.

Middle Ages: Charlemagne had cucumbers grown in his gardens in 9th Century France. They were reportedly introduced into England in the early 14th century, lost, then reintroduced approximately 250 years later. The Spaniards (through the Italian Christopher Columbus) brought cucumbers to Haiti in 1494. In 1535, Jacques Cartier, a French explorer, found "very great cucumbers" grown on the site of what is now Montreal.

Post-Enlightenment: Throughout the 16th century, European trappers, traders, bison hunters, and explorers bartered for the products of American Indian agriculture. The tribes of the Great Plains and the Rocky Mountains learned from the Spanish how to grow European crops. The best farmers on the Great Plains were the Mandan Indians in what is now North and South Dakota.

They obtained cucumbers and watermelons from the Spanish, and added them to the crops they were already growing, including several varieties of corn and beans, pumpkins, squash, and gourd plants. The Iroquois were also growing them when the first Europeans visited them.

In 1630, the Reverend Francis Higginson produced a book called *New England's Plantation* in which, describing a garden on Conant's Island in Boston Harbor known as "The Governor's Garden", he states: "The countrie aboundeth naturally with store of roots of great varietie and good to eat. Our turnips, parsnips, and carrots are here both bigger and sweeter than is ordinary to be found in England. Here are store of pompions, cowcumbers, and other things of that nature which I know not..."

William Wood also published in 1633's New England Prospect (published in England) observations he made in 1629 in America: "The ground affords very good kitchin gardens, for Turneps, Parsnips, Carrots, Radishes, and Pompions, Muskmillons, Isquoter-squashes, coucumbars, Onyons, and whatever grows well in England grows as well there, many things being better and larger."

In the later 17th century, a prejudice developed against uncooked vegetables and fruits. A number of articles in contemporary health publications stated that uncooked plants brought on summer diseases and should be forbidden to children. The cucumber kept this vile reputation for an inordinate period of time: "fit only for consumption by cows", which some believe is why it gained the name, "cowcumber". A copper etching made by Maddalena Bouchard between 1772 and 1793 shows this plant to have smaller, almost bean-shaped fruits, and small yellow flowers. The small form of the cucumber is figured in Herbals of the 16th century, but states, "if hung in a tube while in blossom, the Cucumber will grow to a most surprising length".

Samuel Pepys wrote in his diary on September 22, 1663: "this day Sir W. Batten tells me that Mr. Newhouse is dead of eating cowcumbers, of which the other day I heard of another, I think." In "The Greenstone Door", William Satchell notes that "Te Moanaroa was dead - of a surfeit of cucumbers...", having eaten four of the "prickly" melons. (Chapter XX, The Storm Cloud).

Fredric Hasselquist, in his travels in Asia Minor, Egypt, Cyprus and Palestine in the 18th century, came across the Egyptian or hairy cucumber, *Cucumis chate*. It is said by Hasselquist to be the "queen of cucumbers, refreshing, sweet, solid, and wholesome."

He also states "they still form a great part of the food of the lower-class people in Egypt serving them for meat, drink and physic." George E. Post, in *Hastings' Dictionary of the Bible*, states, "It is longer and more slender than the common cucumber, being often more than a foot long, and sometimes less than an inch thick, and pointed at both ends."

Varieties

In human cultivation, the varieties of cucumbers are classified into three main varieties: "slicing", "pickling", and "burpless".

Slicing: Cucumbers which are grown to be eaten fresh are called slicing cucumbers. They are mainly eaten in the unripe, green form. The ripe, yellow form normally becomes bitter and sour. Slicers grown commercially for the North American market are generally longer, smoother, more uniform in colour, and have a much tougher skin. Slicers in other countries are smaller and have a thinner, more delicate skin. Smaller slicing cucumbers can also be pickled.

Pickling: Cucumbers can be pickled for flavour and longer shelf-life. Although any cucumber can be pickled, commercial pickles are made from cucumbers specially bred for uniformity of length-to-diametre ratio and lack of voids in the flesh. Those cucumbers intended for pickling, called picklers, grow to about 7 cms (3 in) to 10 cms (4 in) long and 2.5 centimetres (1 in) wide.

As compared to slicers, picklers tend to be shorter, thicker, less regularly shaped, and have bumpy skin with tiny white or black-dotted spines. They are never waxed. Colour can vary from creamy yellow to pale or dark green. Pickling cucumbers are sometimes sold fresh as "Kirby" or "Liberty" cucumbers. The pickling process removes or degrades much of the nutrient content, especially that of vitamin C. Pickled cucumbers are soaked in brine or a combination of vinegar and brine, although not vinegar alone, often along with various spices. Pickled cucumbers are called "pickles" in the US or "gherkins" or "wallies" in the UK, the latter name being more common in the north of England, where it refers to the large vinegar-pickled cucumbers commonly sold in fish and chip shops. (Although the gherkin is of the same species as the cucumber, it is of a completely different cultivar.)

English

English cucumbers are sweeter and have a thinner skin than other varieties of cucumber, and are reputed to be easy to digest and to have a pleasant taste. They can grow as long as 2 feet (0.61 m). They are nearly seedless, and have a delicate skin which is pleasant to eat. Most commonly grown in greenhouses, these parthenocarpic cucumbers are often found in grocery markets, shrink-wrapped in plastic. They are sometimes marketed as seedless or burpless, because the seeds and skin of other varieties of cucumbers are said to give some people gas.

Several varietals exist and are sold commercially:

- Lebanese cucumbers are small, smooth-skinned and mild. Like the English cucumber, Lebanese cucumbers are nearly seedless.
- East Asian cucumbers are mild, slender, deep green, and have a bumpy, ridged skin. They can be used for slicing, salads, pickling, etc., and are available year-round. They are usually burpless as well.

- Armenian cucumbers (also known as yard long) have very long, ribbed fruit with a thin skin that does not require peeling, but are actually an immature melon. This is the variety sold in Middle Eastern markets as "pickled wild cucumber". In North America, the term "wild cucumber" refers to manroot.
- Persian cucumber, better known as mini seedless cucumbers, are available from Canada during the summer, and all year-round in the US. Its popularity is increasing 30 to 40% a year. Easy to cut, it is on average 5-8 in. long. Vines are parthenocarpic, requiring no pollinators for fruit set.
- Beit Alpha cucumbers are small, sweet parthenocarpic cucumbers adapted to the dry climate of the Middle East.
- Apple cucumbers are short, round cucumbers grown in New Zealand and parts of Europe, known for their light yellow-green colour and mildly sweet flavour. When mature, the fruit may grow tiny spines, and contains numerous edible green seeds. The fruit is usually eaten raw, with skin.
- *Dosakai* is a yellow cucumber available in parts of India. These fruits are generally spherical in shape. It is commonly cooked as curry, added in *sambar* or soup, *daal* and also in making *dosa-aavakaaya* (Indian pickle) and chutney; it is also grown and available through farms in Central California.
- *Kekiri* is a smooth skinned cucumber, relatively hard, and not used for salads. It is cooked as spicy curry. It is found in dry zone of Sri Lanka. It becomes orange coloured when the fruit is matured.

Taste

There appears to be variability in the human olfactory response to cucumbers, with the majority of people reporting a mild, almost watery flavour or a light melon taste, while a small but vocal minority report a highly repugnant taste, some say almost perfume-like.

There are several ways to remove bitterness from cucumbers. One way used in India, includes slicing off the ends of a cucumber, sprinkling some salt, and rubbing the now-exposed ends of said cucumber with the sliced-off ends until it appears to froth.

Storage

Cucumbers can best be stored sealed in shrink foil. Sealing a cucumber increases its shelf life considerably.

In the News

In May 2011, cucumbers infected with *E. coli* were claimed to have caused the deaths of at least ten people, leading to some retailers withdrawing cucumbers from sale in Germany, Austria and the Czech Republic. The cucumbers were initially thought to have been sourced from Spain, however subsequent test failed to show contamination in the imported Spanish cucumbers which led to the Spanish government demanding compensation for Spanish farmers who had been forced to destroy huge quantities of cucumbers.After the outbreak, the World Health Organisation stated that it was a completely new strain of the bacteria involved.

Muskmelon

Muskmelon (*Cucumis melo*) is a species of melon that has been developed into many cultivated varieties. These include smooth skinned varieties such as honeydew, crenshaw and casaba, and different netted cultivars (cantaloupe, Persian melon and Santa Claus or Christmas melon). The Armenian cucumber is also a variety of muskmelon, but its shape, taste, and culinary uses more closely resemble those of a cucumber. The large number of cultivars in this species approaches that found in wild cabbage, though morphological variation is not as extensive. It is a fruit of a type called pepo. Muskmelon is native to Persia (Iran), Anatolia, Armenia, and adjacent areas on the west and the east which is believed to be their centre of origin and development, with a secondary centre including the northwest provinces of India and Afghanistan. Although truly wild forms of *C. melo* have not been found, several related wild species have been noted in those regions.

Genetics

The genome of *Cucumis melo* L. was sequenced in 2012.

Nutrition

Cantaloupe melons are a good source of potassium, Vitamin A, and folate.

Uses

In addition to their consumption when fresh, melons are sometimes dried. Other varieties are cooked, or grown for their seeds, which are processed to produce melon oil. Still other varieties are grown only for their pleasant fragrance. The Japanese liqueur Midori is flavoured with muskmelon.

Trichosanthes Dioica

Trichosanthes dioica is also known as the pointed gourd, parwal/ parval (from Hindi), or potol (from Assamese, Oriya or Bengali and "Paror" in Maithili). Colloquially, in India, it is often called *green potato*. It is widely cultivated in the eastern part of India, particularly in Orissa, Bengal, Assam, Bihar, and Uttar Pradesh. It is a good source of carbohydrates, vitamin A, and vitamin C. It also contains major nutrients and trace elements (magnesium, potassium, copper, sulfur, and chlorine) which are needed in small quantities, for playing essential roles in human physiology.

It is a vine plant, similar to cucumber and squash, though unlike those it is perennial. It is a dioecious (male and female plants) vine (creeper) plant with heart-shaped leaves (cordate) and is grown on a trellis. The fruits are green with white or no stripes. Size can vary from small and round to thick and long — 2 to 6 inches (5 to 15 cm). It thrives well under a hot to moderately warm and humid climate. The plant remains dormant during the winter season and prefers a fertile, well-drained sandy loam soil due to its susceptibility to water-logging. It is used as ingredients of soup, stew, curry, sweet, or eaten fried and as *potoler dorma* or *dolma* (dolma) with fish, roe or meat stuffing.

Pumpkin

A pumpkin is a gourd-like squash of the genus *Cucurbita* and the family Cucurbitaceae (which also includes gourds). It commonly refers to cultivars of any one of the species *Cucurbita pepo*, *Cucurbita mixta*, *Cucurbita maxima*, and *Cucurbita moschata*, and is native to North America. They typically have a thick, orange or yellow shell, creased from the stem to the bottom, containing the seeds and pulp.

Pumpkins are widely grown for commercial use, and are used both in food and recreation. Pumpkin pie, for instance, is a traditional part of Thanksgiving meals in the United States, and pumpkins are frequently carved as decorations around Halloween. A pumpkin that has a little face carved in it and hollowed out and decorated with candles inside is known as a jack o'lantern; these are often used at Halloween, for example, to decorate windows. In Australian English, the name 'pumpkin' generally refers to the broader category called winter squash in North America.

Description

The word pumpkin originates from the word *pepon*, which is Greek for "large melon". The French adapted this word to *pompon*, which the British changed to *pumpion* and later American colonists changed that to the word we use today, "pumpkin". The origin of pumpkins is not definitively known, although they are thought to have originated in North America. The oldest evidence, pumpkin-related seeds dating between 7000 and 5500 BC, were found in Mexico. Pumpkins are a squash-like fruit that range in size from less than 1 pound (0.45 kilograms) to over 1,000 pounds (450 kilograms).

Since some squash share the same botanical classifications as pumpkins, the names are frequently used interchangeably. In general, pumpkin stems are more rigid, prickly, and angular (with an approximate five-degree angle) than squash stems, which are generally softer, more rounded, and more flared where joined to the fruit.

Pumpkins generally weigh 9–18 lbs (4–8 kg) with the largest (of the species *C. maxima*) capable of reaching a weight of over 75 lbs (34 kg). The pumpkin varies greatly in shape, ranging from oblate to oblong. The rind is smooth and usually lightly ribbed. Although pumpkins are usually orange or yellow, some fruits are dark green, pale green, orange-yellow, white, red and gray.

Pumpkins are monoecious, having both male and female flowers on the same plant. The female flower is distinguished by the small ovary at the base of the petals. These bright and colourful flowers have extremely short life spans and may only open for as short a time as one day. The colour of pumpkins is derived from the orange pigments abundant in them. The main nutrients are lutein and both alpha and beta carotene, the latter of which generates vitamin A in the body.

Taxonomy

Pumpkin is the fruit of the species *Cucurbita pepo* or *Cucurbita mixta*. It can refer to a specific variety of the species *Cucurbita maxima* or *Cucurbita moschata*, which are all of the genus *Cucurbita* and the family Cucurbitaceae.

Distribution and Habitation

Pumpkins are grown all around the world for a variety of reasons ranging from agricultural purposes (such as animal feed) to commercial and ornamental sales. Of the seven continents, only Antarctica is unable to produce pumpkins; the biggest international producers of pumpkins include the United States, Canada, Mexico, India, and China. The traditional American pumpkin is the Connecticut Field variety.

Ecology

Cultivation in the United States: As one of the most popular crops in the United States, 1.5 billion pounds (680,000,000 kilograms) of pumpkins are produced each year. The top pumpkin-producing states include Illinois, Indiana, Ohio, Pennsylvania, and California.

According to the Illinois Department of Agriculture, 95% of the US crop intended for processing is grown in Illinois. Nestlé produces 85% of the processed pumpkin in the United States. In the fall of 2009, rain in Illinois devastated the Nestlé crop, resulting in a shortage affecting the entire country during the Thanksgiving holiday season.

Pumpkins are a warm-weather crop that is usually planted in early July. The specific conditions necessary for growing pumpkins require that soil temperatures three inches (7.6 cm) deep are at least 60 °F (15.5 °C) and soil that holds water well. Pumpkin crops may suffer if there is a lack of water or because of cold temperatures (in this case, below 65 °F (18.3 °C); frost can be detrimental), and sandy soil with poor water retention or poorly drained soils that become waterlogged after heavy rain. Pumpkins are, however, rather hardy, and even if many leaves and portions of the vine are removed or damaged, the plant can very quickly re-grow secondary vines to replace what was removed.

Pumpkins produce both a male and female flower; honeybees play a significant role in fertilization. Pumpkins have historically been pollinated by the native squash bee *Peponapis pruinosa*, but this bee has declined, probably due to pesticide sensitivity, and today most commercial plantings are pollinated by honeybees. One hive per acre (4,000 m^2 per hive) is recommended by the US Department of Agriculture. If there are inadequate bees for pollination, gardeners often have to hand pollinate. Inadequately pollinated pumpkins usually start growing but abort before full development. An opportunistic fungus is also sometimes blamed for abortions.

Giant Pumpkins

The largest pumpkins are *Cucurbita maxima*. They were cultivated from the hubbard squash genotype, crossed with kabocha-pumpkin types by enthusiast farmers through intermittent effort since the early 19th century. As such germplasm is commercially provocative, a US legal right was granted for the rounder phenotypes, levying them as constituting a variety, with the appellation Atlantic Giant. Eventually, this phenotype graduated back into the public domain, except now it had the name Atlantic Giant on its record.

Weigh-off competitions for giant pumpkins are a popular festival activity. The world record held at 460 pounds (208.7 kilograms) until 1981, when Howard Dill (of Nova Scotia) broke the record with a pumpkin near 500 pounds (227 kilograms). Dill patented the seeds used to grow this giant pumpkin, deeming them Dill's Atlantic Giant seeds, and drawing growers from around the world. Dill is credited for all of the giant pumpkins today, most of which are borne from crossing and re-crossing his patented seed with other varieties. By 1994, the Giant pumpkin crossed the 1,000-pound (453.6-kilogram) mark. The current world record holder is Chris Stevens's 1,810-pound Atlantic Giant pumpkin, which in October 2010 surpassed Christy Harp's previous 2009 record of 1,725 pounds.

Extract

East China Normal University research on type-1 diabetic rats, published in July 2007, suggests that chemical compounds found in pumpkin promote regeneration of damaged pancreatic cells, resulting in increased bloodstream insulin levels. According to the research team leader, pumpkin extract may be "a very good product for pre-diabetic people, as well as those who already have diabetes," possibly reducing or eliminating the need for insulin injections for some type-1 diabetics. It is unknown whether pumpkin extract has any effect on diabetes mellitus type 2, as it was not the subject of the study.

Seeds

Pumpkin seeds, also known as pepitas, are small, flat, green, edible seeds. Most pumpkin seeds are covered by a white husk, although some pumpkin varieties produce seeds without them. Pumpkin seeds are a popular snack that can be found hulled or semi-hulled at most grocery stores. However, roasting pumpkin seeds (usually scooped out of jack-o-lanterns) is a popular Halloween treat. Pumpkin seeds have many health benefits, as they are a good source of protein, zinc, and other vitamins, and they are even said to lower cholesterol. One gram of pumpkin seed protein contains as much tryptophan as a full glass of milk. Pumpkin seeds are a good source of magnesium, manganese, phosphorus, and phytosterols.

Pumpkin Seed Oil

Pumpkin seed oil is a thick, green-red oil that is produced from roasted pumpkin seeds. When used for cooking or as a salad dressing, pumpkin seed oil is generally mixed with other oils because of its robust flavour. It is used in cooking in central and eastern Europe.

It is considered a delicacy in Austria, where a little is often added in traditional local cuisine on pumpkin soup and on potato salad. Some restaurants in Vienna even add a few drops on vanilla ice cream. Long believed to be a folk remedy for prostate problems, it has been claimed to combat benign prostatic hyperplasia. Pumpkin seed oil contains essential fatty acids that help maintain healthy blood vessels, nerves and tissues.

Other Uses

Canned pumpkin is often recommended by veterinarians as a dietary supplement for dogs and cats that are experiencing certain digestive ailments such as constipation, diarrhea, or hairballs. The high fiber content helps to aid proper digestion. Raw pumpkin can be fed to poultry, as a supplement to regular feed, during the winter to help maintain egg production, which usually drops off during the cold months. The medicinal properties of pumpkin include anti-diabetic, antioxidant, anti-carcinogenic, and anti-inflammatory.

Activities Involving Pumpkins

Halloween: Pumpkins are commonly carved into decorative lanterns called jack-o'-lanterns for the Halloween season in North America. Throughout Britain and Ireland, there is a long tradition of carving lanterns from vegetables, particularly the turnip, mangelwurzel, or swede.

The turnip has traditionally been used in Ireland and Scotland at Halloween, but immigrants to North America used the native pumpkin, which are both readily available and much larger – making them easier to carve than turnips. Not until 1837, does *jack-o'-lantern* appear as a term for a carved vegetable lantern, and the carved pumpkin lantern association with Halloween is recorded in 1866.

In the United States, the carved pumpkin was first associated with the harvest season in general, long before it became an emblem of Halloween. In 1900, an article on Thanksgiving entertaining recommended a lit jack-o'-lantern as part of the festivities that encourage kids and families to join together to make their own jack-o-lanterns.

Chucking: Pumpkin chucking is a competitive activity in which teams build various mechanical devices designed to throw a pumpkin as far as possible. Catapults, trebuchets, ballistas and air cannons are the most common mechanisms. Some pumpkin chuckers breed and grow special varieties of pumpkin under specialised conditions to improve the pumpkin's chances of surviving a throw.

Pumpkin Festivals and Competitions: Pumpkin growers often compete to see whose pumpkins are the most massive. Festivals are often dedicated to the pumpkin and these competitions. The Ohio towns of Barnesville and Circleville each hold a festival every year, the Barnesville Pumpkin Festival and the Circleville Pumpkin Show respectively.

The town of Half Moon Bay, California, holds an annual Art and Pumpkin Festival, drawing over 250,000 visitors each year and including the World Champion Pumpkin Weigh-Off. Farmers from all over the US compete to determine who can grow the heaviest pumpkin. The winning pumpkin regularly tops the scale at more than 1500 pounds. Leonardo Urena, from Napa, California, grew the winner of the 2011 Weigh-Off with a 1,704-pound Atlantic Giant, setting a new California State record. The world record pumpkin in 2009 was 1725 pounds, grown by Cristy Harp of Ohio.

The town of Morton, Illinois, the self-declared pumpkin capital of the world, has held a Pumpkin Festival since 1966. The town, where Nestlé's pumpkin packing plant is located (and where 90% of canned pumpkins eaten in the US are processed), held for several years a record for the number of carved and lit pumpkins in one place, before losing it to Boston, Massachusetts, in 2006. A large contributor of pumpkins to the Keene Pumpkin Fest in New Hampshire is local Keene State College, which hosts an event called Pumpkin Lobotomy on its main quadrangle. Usually held the day before the festival itself, Pumpkin Lobotomy has the air of a large party, with the school providing pumpkins and carving instruments alike (though some students prefer to use their own) and music provided by college radio station WKNH.

Ireland's only Pumpkin Festival takes place each year in Virginia, County Cavan to find Ireland's biggest pumpkins. This year the biggest pumpkin topped 1300 pounds. The event takes place over a holiday weekend, along with other entertainment and festive parades. The city of Elk Grove, California, has held an annual Pumpkin Festival since 1995.

International Pumpkin Day is observed in various countries around the world on the second Saturday of October. While celebrations take many forms, potluck meals and beer centred around pumpkins are often consumed. It was created in Chicago in 2009.

Folklore and Fiction

There seems to be a strong connection in folklore and popular culture between pumpkins and the supernatural. Famous examples include the following:

Folklore:

- A commonplace motif of people being turned into pumpkins by witches.
- The jack-o-lantern custom discussed above, which connects to Halloween lore about warding off demons.

Fiction:

- In the folk tale *Cinderella*, the fairy godmother turns a pumpkin into a carriage, but at midnight it reverts back into a pumpkin.
- Linus' belief in the Great Pumpkin in Charles M. Schulz's comic strip *Peanuts.*
- Juice from a pumpkin has magical effects in the short story "Pumpkin Juice" by R. L. Stine.
- The *Harry Potter* novels, in which pumpkin juice as a favourite drink of the students of Hogwart's School of Witchcraft and Wizardry is a recurring element.
- The pumpkin hurled by the "Headless Horseman" in Washington Irving's *The Legend of Sleepy Hollow.*
- Jack Pumpkinhead, a character in the Oz books of L. Frank Baum, with a pumpkin for a head on a wooden body, brought to life in the second book.
- In Tim Burton's *The Nightmare Before Christmas,* the main character, Jack Skellington, is "the Pumpkin King".
- Precious Ramotswe, the fictional detective from Botswana in *The No. 1 Ladies' Detective Agency* series of novels by Scottish author Alexander McCall Smith, often cooks and eats pumpkin.
- In a short fiction by Nathaniel Hawthorne, *Feathertop* from 1852, a witch turns a scarecrow with a "pumpkinhead" into a man.

Squash (Plant)

Squashes generally refer to four species of the genus *Cucurbita*, also called marrows (mainly in British English). These species include *C. maxima* (hubbard squash, buttercup squash, some varieties of prize pumpkins, such as Big Max), *C. mixta* (cushaw squash), *C. moschata* (butternut squash), and *C. pepo* (most pumpkins, acorn squash, summer squash, zucchini).

In North America, squash is loosely grouped into summer squash or winter squash, depending on whether they are harvested as immature fruit (summer squash) or mature fruit (autumn squash or winter squash). Gourds are from the same family as squashes. Well known types of squash include the pumpkin and zucchini. Giant squash are derived from *Cucurbita maxima* and are routinely grown to weights nearing those of giant pumpkins. For more details, refer to list of gourds and squashes.

Cultivation

Archaeological evidence suggests squash may have been first cultivated in Mesoamerica some 8,000 to 10,000 years ago, and may have been independently cultivated elsewhere at a later date. Squash was one of the "Three Sisters" planted by Native Americans. The Three Sisters were the three main native crop plants: maize (corn), beans, and squash. These were usually planted together, with the cornstalk providing support for the climbing beans, and shade for the squash. The squash vines provided ground cover to limit weeds. Weeds can be detrimental to the growing conditions of the squash. The beans provided nitrogen fixing for all three crops.

Summer squashes, including zucchini (also known as courgette), pattypan and yellow crookneck are harvested during the growing season, while the skin is still soft and the fruit rather small; they are eaten almost immediately and require little to no cooking. Winter squashes (such as butternut, Hubbard, buttercup, ambercup, acorn, spaghetti squash and pumpkin) are harvested at maturity, generally the end of summer, cured to further harden the skin, and stored in a cool place for eating later.

They generally require longer cooking time than summer squashes. (Note: Although the term "winter squash" is used here to differentiate from "summer squash", it is also commonly used as a synonym for *Cucurbita maxima.*) The squash fruit is classified as a pepo by botanists, which is a special type of berry with a thick outer wall or rind formed from hypanthium tissue fused to the exocarp; the fleshy interior is composed of mesocarp and endocarp.

The pepo, derived from an inferior ovary, is characteristic of the squash family (Cucurbitaceae). In culinary terms, both summer and winter squashes are generally considered as vegetables, even though pumpkin may be used for sweet dishes.

In addition to the fruit, other parts of the plant are edible. Squash seeds can be eaten directly, ground into paste, meal, "nut" butter, even a fine flour, or (particularly for hulless pumpkins) pressed for vegetable oil (e.g. bottle gourd, buffalo gourd, and pumpkin seed oils).

The shoots, leaves, and tendrils can be eaten as greens. The blossoms are an important part of native American cooking, and are also used in many other parts of the world. Both the male and female blossoms can be harvested pre- or mid-flower.

Pollination

As with all other members of the family, the flowers come in pollen-bearing male form, and the ovary-bearing female form, with both forms being present on the plant.

Squash has historically been pollinated by the native North American squash bee *Peponapis pruinosa*, and related species, but this bee and its relatives have declined, probably due to pesticide sensitivity, and most commercial plantings are pollinated by European honey bees today.

One hive per acre (4,000 m^2 per hive) is recommended by the US Department of Agriculture. Gardeners with a shortage of bees often have to hand pollinate. Giant squash grown competitively are usually hand pollinated. Flowers are kept closed before and after pollination to prevent cross-pollination. Inadequately pollinated female squash flowers will usually start growing, but abort before full development. Many gardeners blame various fungal diseases for the aborted fruit, but the fix proves to be better pollination, not fungicide.

Preparation

Though considered a vegetable in cooking, botanically speaking, squash is a fruit (being the receptacle for the plant's seeds).

Squash can be served fresh (in salads) and cooked (squash stuffed with meat, fried squash, baked squash). Small pattypans are good for pickling.

Uses in Art

The squash has been an essential crop in the Andes since the pre-Columbian Era. This pottery was a sacred substance, formed in significant shapes and used to represent important themes. Squash are represented frequently in Moche ceramics.

Luffa

The luffa, loofah, or lufah are tropical and subtropical vines comprising the genus *Luffa*, the only genus of the subtribe Luffinae of the plant family Cucurbitaceae. The fruit of at least two species, *Luffa acutangula* and *Luffa aegyptiaca (Luffa cylindrica)*, is grown, harvested before maturity, and eaten as a vegetable, popular in Asia and Africa.

The ripe, dried fruit is also the source of the loofah or plant sponge.

Luffa species are used as food plants by the larvae of some Lepidoptera species, including *Hypercompe albicornis.*

Parts of the plant are used to create bath or kitchen sponges, a natural jaundice remedy, furniture and even houses. The term is also used to describe synthetic bath tools that serve the same purpose.

Uses

The fruit section of *L. aegyptiaca* may be allowed to mature and used as a bath or kitchen sponge after being processed to remove everything but the network of xylem or fibers. Marketed as *luffa* or *loofah*, the sponge is used like a body scrub.

Its juice is used as a natural remedy for jaundice. The juice is obtained by pounding the bitter luffa and squeezing it through a cloth. Bitter luffa seeds and dry crusts are also available and can be used for the same purpose.

In Maharashtra, India, dodka (Ridge Gourd/luffa) and ghosavala (smooth luffa) are common vegetables prepared with either crushed dried peanuts or with beans. In northern India as well as Pakistan, Turai (thoo-raee) is the common name for Luffa. In China, Indonesia, the Philippines, Manipur, (India) (called Sebot) the luffa or patola is eaten as a green vegetable in various dishes. The luffa is eaten when the fruit is young and the sponge has yet to become tough. It is also known as *Chinese Okra* in Canada.

In Paraguay, panels are made out of luffa combined with other vegetable matter and recycled plastic. These can be used to create furniture and construct houses.

Trichosanthes Cucumerina

'*Trichosanthes cucumerina*' is a tropical or subtropical vine, raised for its strikingly long fruit, used as a vegetable, medicine, and, a lesser known use, crafting didgeridoos. Common names include snake gourd (var. *anguina*), serpent gourd, chichinga, and padwal.

It is known as chichindo in Nepali, chhachindra in Oriya, *chichinga* or *chichinge* in Bengali, *purla* in Sambalpuri language, *potlakaaya* in Telugu, *pathola* in Sinhala, *pudalankaai* in Tamil, *dhunduli* in Assamese, *paduvalakaayi* in Kannada and *padavalanga* in Malayalam.

Formerly, the cultivated form was considered a distinct species, but it is nowadays regarded as a mere variety of the wild ancestor, as they freely interbreed:

- *Trichosanthes cucumerina* var. *anguina* (L.) Haines – cultivated variant
- *Trichosanthes cucumerina* var. *cucumerina* – wild variant

The narrow, soft-skinned fruit can reach 150 cm long. Its soft, bland, somewhat mucilaginous flesh is similar to that of the luffa and the calabash. It is most popular in the cuisine of South Asia and Southeast Asia. The shoots, tendrils, and leaves are also eaten as greens.

Watermelon

Watermelon (*Citrullus lanatus* (Thunb.), family Cucurbitaceae) is a vine-like (scrambler and trailer) flowering plant originally from southern Africa. Its fruit, which is also called *watermelon,* is a special kind referred to by botanists as a pepo, a berry which has a thick rind (exocarp) and fleshy centre (mesocarp and endocarp).

Pepos are derived from an inferior ovary, and are characteristic of the Cucurbitaceae. The watermelon fruit, loosely considered a type of melon – although not in the genus *Cucumis* – has a smooth exterior rind (green, yellow and sometimes white) and a juicy, sweet interior flesh (usually deep red to pink, but sometimes orange, yellow and even green if not ripe).

History

Watermelon is thought to have originated in southern Africa, where it is found growing wild. It reaches maximum genetic diversity there, with sweet, bland and bitter forms. In the 19th century, Alphonse de Candolle claimed the watermelon was indigenous to tropical Africa. Though *Citrullus colocynthis* is often considered to be a wild ancestor of watermelon and is now found native in north and west Africa, it has been suggested on the basis of chloroplast DNA investigations that the cultivated and wild watermelon diverged independently from a common ancestor, possibly *C. ecirrhosus* from Namibia.

Evidence of its cultivation in the Nile Valley was found from the second millennium BC. Watermelon seeds have been found at Twelfth Dynasty sites and in the tomb of Pharaoh Tutankhamun. Watermelon is also mentioned in the Bible as a food eaten by the ancient Israelites while they were in bondage in Egypt.

By the 10th century, watermelons were being cultivated in China, which is today the world's single largest watermelon producer. By the 13th century, Moorish invaders had introduced the fruit to Europe; according to John Mariani's *Dictionary of American Food and Drink*, "watermelon" made its first appearance in an English dictionary in 1615.

Watermelons were grown by Native Americans in the 16th century. Early French explorers found the fruit being cultivated in the Mississippi Valley. Many sources list the watermelon as being introduced in Massachusetts as early as 1629. Southern food historian John Egerton has said he believes African slaves helped introduce the watermelon to the United States. Texas Agricultural Extension horticulturalist Jerry Parsons lists African slaves and European colonists as having distributed watermelons to many areas of the world. Parsons also mentions the crop being farmed by Native Americans in Florida (by 1664) and the Colorado River area (by 1799). Other early watermelon sightings include the Midwestern states (1673), Connecticut (1747) and the Illiana region (1822).

Charles Fredric Andrus, a horticulturist at the USDA Vegetable Breeding Laboratory in Charleston, South Carolina, set out to produce a disease-resistant and wilt-resistant watermelon. The result, in 1954, was "that gray melon from Charleston". Its oblong shape and hard rind made it easy to stack and ship. Its adaptability meant it could be grown over a wide geographical area. It produced high yields and was resistant to the most serious watermelon diseases: anthracnose and fusarium wilt.

Today, farmers in approximately 44 states in the US grow watermelon commercially, and almost all these varieties have some 'Charleston Gray' in their lineage. Georgia, Florida, Texas, California and Arizona are the US's largest watermelon producers. This now-common watermelon is often large enough that groceries often sell half or quarter melons. Some smaller, spherical varieties of watermelon, both red- and yellow-fleshed, are sometimes called "icebox melons".

Culture

For commercial plantings, one beehive per acre (4,000 m^2 per hive) is the minimum recommendation by the US Department of Agriculture for pollination of conventional, seeded varieties. Because seedless hybrids have sterile pollen, pollinizer rows of varieties with viable pollen must also be planted.

Since the supply of viable pollen is reduced and pollination is much more critical in producing the seedless variety, the recommended number of hives per acre, or pollinator density, increases to three hives per acre (1,300 m^2 per hive). Watermelons have a longer growing period than other garden plants and can often take up to 85 days of growing to mature.

Varieties

The more than 1200 varieties of watermelon range in weight from less than one to more than 200 pounds; the flesh can be red, orange, yellow or white.

- The 'Carolina Cross' produced the current world record watermelon weighing 262 pounds (119 kg). It has green skin, red flesh and commonly produces fruit between 65 and 150 pounds (29 and 68 kg). It takes about 90 days from planting to harvest.
- The 'Yellow Crimson' has a yellow-coloured flesh. It has been described as sweeter and more honey-flavoured than the more popular red-flesh watermelon.
- The 'Orangeglo' has a very sweet orange flesh, and is a large, oblong fruit weighing 9–14 kg (20–30 pounds). It has a light green rind with jagged dark green stripes. It takes about 90–100 days from planting to harvest.
- The 'Moon and Stars' variety was created in 1926. The rind is purple/black and has many small, yellow circles (stars) and one or two large, yellow circles (moon). The melon weighs 9–23 kg (20–50 pounds). The flesh is pink or red and has brown seeds. The foliage is also spotted. The time from planting to harvest is about 90 days.
- The 'Cream of Saskatchewan' consists of small, round fruits around 25 cm (10 inches) in diametre. It has a quite thin, light green with dark green striped rind, with sweet white flesh and black seeds. It can grow well in cool climates. It was originally brought to Saskatchewan, Canada, by Russian immigrants. The melon takes 80–85 days from planting to harvest.
- The 'Melitopolski' has small, round fruits roughly 28–30 cm (11–12 inches) in diametre. It is an early ripening variety that originated from the Volga River region of Russia, an area known for cultivation of watermelons. The Melitopolski watermelons are seen piled high by vendors in Moscow in summer. This variety takes around 95 days from planting to harvest.

- The 'Densuke' watermelon has round fruit up to 25 lb (11 kg). The rind is black with no stripes or spots. It is grown only on the island of Hokkaido, Japan, where up to 10,000 watermelons are produced every year. In June 2008, one of the first harvested watermelons was sold at an auction for 650,000 yen (US$ 6,300), making it the most expensive watermelon ever sold. The average selling price is generally around 25,000 yen ($ 250).

In Japan, farmers of the Zentsuji region found a way to grow cubic watermelons, by growing the fruits in glass boxes and letting them naturally assume the shape of the receptacle. The cubic shape was originally designed to make the melons easier to stack and store, but the cubic watermelons are often more than double the price of normal ones, and much of their appeal to consumers is in their novelty. Pyramid-shaped watermelons have also been developed and any polyhedral shape may potentially also be used.

Cultural References

- In Vietnamese culture, watermelon seeds are consumed during the Vietnamese New Year's holiday, *T¿t*, as a snack.
- Stereotypical caricatures may depict African Americans as being inordinately fond of watermelon, to the point where some African Americans do not want to be seen in public eating watermelon.
- The Oklahoma State Senate passed a bill on 17 April 2007 declaring watermelon as the official state vegetable, with some controversy surrounding whether a watermelon is a vegetable or a fruit.
- The citrulline in watermelon (especially in the rind) is a known stimulator of nitric oxide. Nitric oxide is thought to relax and expand blood vessels, much like the erectile dysfunction drug Viagra, and may even increase libido.
- Fans of the Saskatchewan Roughriders of the CFL started a tradition of hollowing out a watermelon and wearing it as a makeshift football helmet (the colour of the Roughriders is green). During the 2009 Grey Cup in Calgary (between the Montreal Alouettes and the Roughriders), thousands of watermelons had to be imported to Calgary supermarkets to prevent a shortage being caused by Rider fans.
- The town of Chinchilla in Queensland, Australia, holds a biannual festival celebrating all things melon.

- The ten-lined June beetle is often affectionately referred to as a watermelon beetle, due to the green, striped pattern on its back.
- It is the symbol of the Turkish city, Diyarbakýr.

Dioscoreaceae

Dioscoreaceae is a family of monocotyledonous flowering plants, with about 750 species in eight or nine genera. The best-known member of the family is the Yam (*Dioscorea*).

The APG system (1998) and APG II system (2003) both place it in the order Dioscoreales, in the clade monocots. However, the circumscription changed in the APG II system, with the 2003 system expanded to include the plants that in the 1998 system were treated in the families Taccaceae and Trichopodaceae.

Genera

Dioscoreaceae (*sensu stricto*)

- *Borderea*
- *Dioscorea*
- *Epipetrum*
- *Rajania*
- *Stenomeris*
- *Tamus* (often included in *Dioscorea*)

(Taccaceae)

- *Tacca*

(Trichopodaceae)

- *Avetra*
- *Trichopus*

Yam (Vegetable)

Yam is the common name for some species in the genus *Dioscorea* (family Dioscoreaceae). These are perennial herbaceous vines cultivated for the consumption of their starchy tubers in Africa, Asia, Latin America, the Caribbean and Oceania. There are many cultivars of yam. Although the sweet potato (*Ipomoea batatas*) has traditionally been referred to as a *yam* in parts of the United States and Canada, it is not part of the Dioscoreaceae family.

Yam is a versatile vegetable. It can be barbecued; roasted; fried; grilled; boiled; baked; smoked and when grated it is processed into a dessert recipe. Yams are the staple crop of the Igbo people of Nigeria, in their language it is known as *ji*, and they commemorate it by having yam festivals known as *Iri-ji* or *Iwa-Ji* depending on the dialect.

Yams are a primary agricultural and culturally important commodity in West Africa, where over 95 percent of world's yam crop is harvested. Yams are still important for survival in these regions. Some varieties of these tubers can be stored up to six months without refrigeration, which makes them a valuable resource for the yearly period of food scarcity at the beginning of the wet season. Yam cultivars are also cropped in other humid tropical countries.

Yam tubers can grow up to 1.5 metres (4.9 feet) in length and weigh up to 70 kilograms (154 lb) and 3 to 6 inches high. The vegetable has a rough skin which is difficult to peel, but which softens after heating. The skins vary in colour from dark brown to light pink. The majority of the vegetable is composed of a much softer substance known as the "meat". This substance ranges in colour from white or yellow to purple or pink in mature yams.

Due to their abundance and importance to survival, the yam was highly regarded in Nigerian ceremonial culture and is part of many West African ceremonies. Certain species of yams are a competing phytochemical source. Yams are used as food plants by the larvae of some Lepidoptera species including *Palpifer sordida*.

Major Cultivated Species

There are many cultivars of yam throughout the humid tropics. The most economically important are discussed below.

Dioscorea Rotundata *and* D. Cayenensis

Dioscorea rotundata, the "white yam", and *Dioscorea cayenensis*, the "yellow yam", are native to Africa. They are the most important cultivated yams. In the past they were considered two separate species but most taxonomists now regard them as the same species. There are over 200 cultivated varieties between them.

White yam's tuber is roughly cylindrical in shape, the skin is smooth and brown and the flesh usually white and firm. Yellow yam is named after its yellow flesh, a colour caused by the presence of carotenoids. It looks similar to the white yam in outer appearance; its tuber skin is usually a bit firmer and less extensively grooved. The yellow yam has a longer period of vegetation and a shorter dormancy than white yam.

The Kokoro variety is important in making dried yam chips.

They are large plants; the vines can be as long as 10 to 12 metres (33 to 39 ft). The tubers most often weigh about 2.5 to 5 kilograms (5.5 to 11 lb) each but can weigh as much as 25 kilograms (55 lb). After 7 to 12 months growth the tubers are harvested. In Africa most are pounded into a paste to make the traditional dish of "pounded yam" (Kay 1987).

D. Alata: *Dioscorea alata,* called "water yam", "winged yam" and "purple yam" (not to be confused with the Okinawan purple "yam", which is a sweet potato), was first cultivated in Southeast Asia. Although not grown in the same quantities as the African yams, it has the largest distribution world-wide of any cultivated yam, being grown in Asia, the Pacific islands, Africa, and the West Indies (Mignouna, 2003). Even in Africa, the popularity of water yam is second only to white yam. The tuber shape is generally cylindrical, but can vary. Tuber flesh is white and watery in texture.

In the United States, water yam has become an invasive species in some southern states. In the Philippines it is known as "ube" and is used as an ingredient in many sweet desserts. In Indonesia it is known as "ubi". In Vietnam, it is called *khoai mú* and is used mainly as an ingredient for soup. In India, it is known as *ratalu* or *violet yam.* In Hawaii it is known as *uhi.*

Uhi was brought to Hawaii by the early Polynesian settlers and became a major crop in the 19th century when the tubers were sold to visiting ships as an easily stored food supply for their voyages (White, 2003).

D. Opposite: *Dioscorea opposita,* "Chinese yam", is native to China. The Chinese yam plant is somewhat smaller than the African, with the vines about 3 metres (10 feet) long. It is tolerant to frost and can be grown in much cooler conditions than other yams. It is now grown in China, Korea, and Japan.

It was introduced to Europe in the 19th century when the potato crop there was falling victim to disease, and is still grown in France for the Asian food market.

The tubers are harvested after about 6 months of growth. Some are eaten right after harvesting and some are used as ingredients for other dishes, including noodles, and for traditional medicines (Kay, 1987).

D. Bulbifera: *Dioscorea bulbifera,* the "air potato", is found in both Africa and Asia, with slight differences between those found in each place. It is a large vine, 6 metres (20 ft) or more in length. It produces tubers; however the bulbils which grow at the base of its leaves are the more important food product. They are about the size of potatoes (hence the name "air potato"), weighing from 0.5 to 2 kilograms (1.1 to 4.4 lb).

Some varieties can be eaten raw while some require soaking or boiling for detoxification before eating. It is not grown much commercially since the flavour of other yams is preferred by most people. However it is popular in home vegetable gardens because it produces a crop after only four months of growth and continues producing for the life of the vine, as long as two years. Also the bulbils are easy to harvest and cook (Kay, 1987).

In 1905 the air potato was introduced to Florida and has since become an invasive species in much of the state. Its rapid growth crowds out native vegetation and is very difficult to remove since it can grow back from the tubers, and new vines can grow from the bulbils even after being cut down or burned (Schultz, 1993).

D. Esculenta: *Dioscorea esculenta*, the "lesser yam", was one of the first yam species cultivated. It is native to Southeast Asia and is the third most commonly cultivated species there, although it is cultivated very little in other parts of the world. Its vines seldom reach more than 3 metres (10 feet) in length and the tubers are fairly small in most varieties.

The tubers are eaten baked, boiled, or fried much like potatoes. Because of the small size of the tubers, mechanical cultivation is possible; which, along with its easy preparation and good flavour, could help the lesser yam to become more popular in the future (Kay, 1987).

D. Dumetorum: *Dioscorea dumetorum*, the "bitter yam", is popular as a vegetable in parts of West Africa; one reason being that their cultivation requires less labour than other yams. The wild forms are very toxic and are sometimes used to poison animals when mixed with bait. It is said that they have also been used for criminal purposes (Kay, 1987).

D. Trifida: *Dioscorea trifida*, the "cush-cush yam", is native to the Guyana region of South America and is the most important cultivated New World yam. Since they originated in tropical rain forest conditions their growth cycle is less related to seasonal changes than other yams. Because of their relative ease of cultivation and their good flavour they are considered to have a great potential for increased production (Kay, 1987).

Production

Yams are farmed on about 5 million hectares in about 47 countries in tropical and subtropical regions of the world. Yam crop begins when whole seed tubers or tuber portions are planted into mounds or ridges, at the beginning of the rainy season. The crop yield depends on how and where the setts are planted, sizes of mounds, interplant spacing, provision of stakes for the resultant plants, yam species, and tuber sizes desired at harvest.

Small-scale farmers in West and Central Africa often intercrop yams with cereals and vegetables. The seed yams are perishable and bulky to transport. Farmers who do not buy new seed yams, usually set aside up to 30 percent of their harvest for planting the next year. Yam crops face pressure from a range of insect pests, fungal and viral diseases, as well as nematodes.

Yam typically grows for six to ten months and is dormant for two to four months, depending on the species. The growth and dormant phases correspond respectively to the wet season and the dry season. For maximum yield the yam requires a humid tropical environment, with an annual rainfall of over 1,500 mm distributed uniformly throughout the growing season. White, yellow and water yams typically produce a single large tuber per year, generally weighing 5 to 10 kg. Yam production requires high labour. The crop has low yield per hectare compared to crops such as cassava (manioc) or sweet potato. It is not an efficient food staple given the relatively large amount of planting material that is required and its long growing season. Yam's labour requirement exceeds that of other comparable crops. Yam is also difficult to preserve and store over extended periods of time. The cost per 1,000 calories of yam is four times greater than those of other root and tuber crops.

For these reasons and problems of storing harvested yam, the costs of yam production are high and yam crops are slowly losing ground to *cassava* and other food staples. Despite these high costs and low nutrient density, when compared to other tubers and roots, low technology yam farming produces the highest amount of food calorie and protein annually per hectare per season, on average. Given this nutritional value of yam and its high cultural acceptance in certain parts of Africa, there is an interest in developing knowledge that can improve yam agriculture. In 2010, the world harvested 48.7 million tonnes of yam, 95 percent of which was produced in Africa. The biggest yam harvest, globally, was in 2008, when the world produced 54 million metric tonnes of yam.

The world average annual yield of yams was 10.2 tonnes per hectare in 2010. The most productive yam farms in the world were in Colombia, where natiowide average annual yield was 28.3 tonnes per hectare. These are average national yields; certain farms report yields significantly above 30 tonnes per hectare, for example for yellow yam; other farms report yields less than 1 tonnes per hectare.

Despite the high labour requirements and production costs, consumer demand for yam is very high in certain subregions of Africa, making yam cultivation quite profitable to certain farmers.

Harvesting

Harvesting of yams in west Africa is typically done by hand using sticks, spades or diggers. Wood-based tools are preferred to metallic tools as they are less likely to damage the fragile tubers; however, wood tools need frequent replacement. Yam harvesting is labour-intensive and physically demanding.

For each 2-10 kilogram tuber harvested, it involves standing, bending, squatting, and sometimes sitting on the ground depending the size of mound, size of tuber or depth of tuber penetration. Care must be taken to avoid damage to the tuber, because damaged tubers do not store well and spoil rapidly.

Some farmers use staking and mixed cropping, a practice that complicates harvesting in some cases. In forested areas, tubers grow in areas where other tree roots are present. Harvesting the tuber then involves the additional step of freeing them from other roots. This often causes tuber damage.

Aerial tubers or bulbils are harvested by manual plucking from the vine. Yields may improve and cost of yam production be lower if mechanization were to be developed and adopted. However, current crop production practices and species used pose considerable hurdles to successful mechanization of yam production, particularly for small-scale rural farmers.

Extensive changes in traditional cultivation practices, such as mixed cropping, may be required. Modification of current tuber harvesting equipment is necessary given yam tuber architecture and its different physical properties.

Storage

Roots and tubers such as yam are living organisms. When stored, they continue to respire. The respiration process results in the oxidation of the starch (a polymer of glucose) contained in the cells of the tuber, which converts it into water, carbon dioxide and heat energy. During this transformation of the starch the dry matter of the tuber is reduced.

Amongst the major roots and tubers, properly stored yam is considered to be the least perishable. Successful storage of yams requires:

- avoid bad and damaged yams, use only sound and healthy yams;
- proper curing, if possible combined with fungicide treatment;
- adequate ventilation to remove the heat generated by respiration of the tubers;
- regular inspection during storage and removal of rotting tubers and any sprouts that develop and;
- protection from direct sunlight and rain.

Storing yam at low temperature reduces the respiration rates. However, temperatures below 12°C (53.6°F) cause damage through chilling, causing a breakdown of internal tissues, increasing water loss and yam's susceptibility to decay. The symptoms of chilling injury are not always obvious when the tubers are still in cold storage. The injury becomes noticeable as soon as the tubers are restored to ambient temperatures.

The best temperature to store yams is between 14-16°C (57.2-60.8°F), with high technology controlled humidity and climatic conditions, after a process of curing. Most countries that grow yam as food staple are too poor to afford high technology storage systems. Sprouting rapidly increases a tuber's respiration rates, and accelerates the rate at which its food value decreases.

Certain cultivars of yam store better than others. The easier to store yams are those that are adapted to arid climate, where they tend to stay in dormant low-respiration stage for much longer than yam breeds that are adapted to humid tropical lands where they do not need dormancy. Yellow yam and cush-cush yam, by nature, have much shorter dormancy periods than water yam, white yam or lesser yam. Storage losses for yams are very high in Africa, with insect alone causing over 25 percent harvest loss within 4 months.

Toxicity in Yams

Unlike cassava, most varieties of edible, mature, cultivated yam do not contain toxic compounds. However, there are exceptions. Bitter compounds tend to accumulate in immature tuber tissues of white and yellow yams. These may be polyphenols or tanninlike compounds. Wild forms of bitter yams do contain some toxins that taste bitter, and hence are referred to as bitter yam. Bitter yams are not normally eaten except at times of desperation in poor countries and in times of local food scarcity. They are usually detoxified by soaking in a vessel of salt water, in cold or hot fresh water or in a stream. The bitter compounds in these yams are water soluble alkaloids which, on ingestion, produce severe and distressing symptoms. Severe cases of alkaloid intoxication may prove fatal.

Aerial or potato yams too have antinutritional factors. In Asia, detoxification methods, involving water extraction, fermentation and roasting of the grated tuber are used for bitter cultivars of this yam. The bitter compounds in yams also known locally as *air potato* include diosbulbin and possibly saponins.

These substances are toxic, causing paralysis. Extracts are sometimes used in fishing to immobolize the fish and thus facilitate capture. The community of Zulus use this yam as bait for monkeys, while hunters in Malaysia use it to poison wildlife such as tigers. In Indonesia an extract of *air potato* is used in the preparation of arrow poison.

Nutritional Value

The protein content and quality of roots and tubers is lower than other food staples. Of all roots and tubers, the protein content of yam and potato is the highest, being approximately 2 percent on a fresh weight basis. Yams, with cassava, provide a much greater proportion of the protein intake in Africa, ranging from 6 percent in East and southern Africa to about 16 percent in humid West Africa.

Yam, like other root crops, is not a good source of essential amino acids. It is rich in phenylalanine and threonine but limiting in the sulphur amino-acids, cystine and methionine and in tryptophan.

Yam consuming areas of Africa have a high incidence of kwashiorkor, a serious medical condition in children caused by protein deficiency. Experts emphasize the need to supplement a yam-driven diet with more protein-rich foods in order to support active and healthy growth in infants.

Except for potassium, vitamin B_6 and vitamin C, yam is a food with low nutrient density. Yam provides around 110 calories per 100 grams. Yam is high in vitamins C and B_6, potassium, manganese and dietary fiber while being low in saturated fat and sodium. A product that is high in potassium and low in sodium is likely to produce a good potassium-sodium balance in the human body, and so protects against osteoporosis and heart disease.

Yam products generally have a lower glycemic index than potato products. The African yam (Dioscorea sp) contains thiocyanate which was implicated by one author as potentially important in sickle cell anemia, an effect that remains scientifically unsubstantiated.

Nutritional significance of yam

Yam's importance in the diet of the people in Nigeria and West African countries cannot be overemphasized. It contributes more than 200 calorie per person per day for more than 150 million people in West Africa while servicing as an important source of income to the people. Yam is an attractive crop in poor farms with limited resources. It is rich in starch, and allows many recipes. Yam is also available all year round making it preferable to other unreliable seasonal crops. These characteristics make yam a preferred food and a culturally important food security crop in some sub-Saharan African countries.

Phytochemical Value

The tubers of certain wild yam, a variant of kokoro yam and other species of *Dioscorea*, such as *Dioscorea nipponica*, are a source for the extraction of diosgenin, a steroid sapogenin. The extracted diosgenin is used for the commercial synthesis of cortisone, pregnenolone, progesterone, and other steroid products. Such preparations were used in early combined oral contraceptive pills. The unmodified steroid has estrogenic activity.

Preparation

Africa: Yams of African species must be cooked to be safely eaten, because various natural substances in raw yams can cause illness if consumed. (Excessive skin contact with uncooked yam fluids can cause the skin to itch. If this occurs, a quick cold bath or application of red palm oil to the affected part of the body will stop the itching.) The most common cooking method in Western and Central Africa is by boiling, frying and roasting the yam.

Preparing boiled yam involves cutting the tuber into five or so thick slices, then peeling the skin from each cut piece and boiling the whitish starchy flesh. The older the yam, the smaller the chunks must be cut and the more water it will need to boil. Boiled yam is traditionally consumed with palm oil, but is also commonly served with a sauce such as pepper sauce or palaver sauce. In big cities and border towns in West Africa, fried yam and pepper sauce is a popular street food and has a similar position as french fries and ketchup in Western cuisine. Yam balls (styled after meat balls) have also gained some popularity in contemporary West African cuisine.

Among the Akan of Ghana, boiled yam can be mashed with palm oil into *eto* in a similar manner to the plantain dish *matoke*, and is served with eggs. The boiled yam can also be pounded with a traditional mortar and pestle to create a thick, starchy paste known as *iyan* ("pounded yam") or *fufu* which is eaten with traditional sauces such as egusi and palmnut soup.

Another method of consumption is to leave the raw yam pieces to dry in the sun. When dry, the pieces turn a dark brown colour. These are then milled to create a brown powder known in Nigeria as *elubo*. The powder can be mixed with boiling water to create a thick starchy paste known as amala, which is used in local soups and sauces.

The Philippines: In the Philippines, the purple *ube* species of yam (*Dioscorea alata*), is eaten as a sweetened dessert called "ube halaya", and is also used as an ingredient in another Filipino dessert, *halo-halo*.

Vietnam: In Vietnam, the same purple yam is used for preparing a special type of soup *canh khoai mỡ* or *fatty yam soup*. This involves mashing the yam and cooking it until very well done.

Indonesia: In Indonesia, the same purple yam is used for preparing desserts. This involves mashing the yam and mixing it with coconut milk and sugar.

Japan: An exception to the cooking rule is the Japanese mountain yam (*Dioscorea opposita*), known as *nagaimo* or *yamaimo* (q\‹,) depending on the root shape.

Nagaimo is eaten raw and grated, after only a relatively minimal preparation: the whole tubers are briefly soaked in a vinegar-water solution to neutralize irritant oxalate crystals found in their skin. The raw vegetable is starchy and bland, mucilaginous when grated, and may be eaten plain as a side dish, or added to noodles. Another variety of yam, *Jinenjo*, is used in Japan as an ingredient in soba noodles.

In Japan, the purple yam is popular as lightly deep fried tempura as well as being grilled or boiled. Additionally, the purple yam is a common ingredient of yam ice cream with the signature purple colour.

The West: 'Yam powder' is available in the West from grocers specialising in African products, and may be used in a similar manner to instant mashed potato powder, although preparation is a little more difficult because of the tendency to form lumps.

The 'yam powder' is sprinkled onto a pan containing a small amount of boiling water, and stirred vigorously. The resulting mixture is served with a heated sauce, such as tomato and chili, poured onto it. To avoid lumps forming, the powder and the water are mixed before putting on the heat, then stirred continuously as the mixture is heating.

Cultural Aspects

Nigeria and Ghana: Yam Festival is usually held in the beginning of August at the end of the rainy season. A popular holiday in Ghana, the Yam Festival is so named because yam is the most common food in many African countries. Yams are the first crops to be harvested. People offer yams to gods and ancestors first, before distributing them to the villagers. This is their way of giving thanks to the spirits above them.

New Yam Festival (Igbo in South-East, Nigeria): The New Yam Festival consists of prayers and thanks for the years past. Yam is the main agricultural crop of the Igbos, Idomas, and Tivs. It is the "staple" food of the Igbo people. The New Yam Festival, known as Orureshi in Owukpa in Idoma west and Ima-Ji, Iri-Ji or Iwa Ji in Igbo land is a celebration depicting the prominence of yam in the social and cultural life. The festival is very promiment among south-eastern states and all the major tribes in Benue state, mainly around August.

The celebration is followed by various cultural dance with the display of masquerades from different clans or groups. This usually last to very late night. During the occasion is also a moment of reuniting old friends and family members.

Traditionally, there was "The Feast of the New Yam", which was held every year before the harvest began in many Igbo villages prior to colonialism. This feast was held to honour the earth goddess and the ancestral spirits of the clan. New yams could not be eaten until some had first been offered to these powers.

Elsewhere: Historical records in West Africa and of African yams in Europe date back to the 16th century. Yams were taken to the Americas through precolonial Portuguese and Spanish on the borders of Brazil and Guyana, followed by a dispersion through the Caribbean.

The coming of the yams (one of the numerous versions from Maré) is described in Pene Nengone (Loyalty Islands - New Caledonia). In many societies yams are so important that one can speak of a 'yam civilization'. Growing the tuber is associated with magic; the best ones must be given to the chief or king; there is a series of myths connected to a divine origin; a farmer may gain a lot of prestige by growing the largest or longest yam.

In Tonga, the ancient names of the months of the year, and the names of the days of the moon-month, were all geared towards the growing of yam. People of ancient times worshiped the yam. Olhuala is a type of local yam that is a staple food in the Maldives.

On the Japanese island of Rishiri, yams and yam products are regarded as a folk remedy for the treatment of impotence, possibly because of the vegetable's high vitamin E content, but likely because of its evocation of virile phallic imagery, according to the common folk medicine theory of sympathetic medicine.

Etymology

From Portuguese inhame or Spanish iñame or ñame, which are probably derived from some West African language, such as Fulani, Serer, Wolof, etc., many of which have verbs meaning "to eat", which could have been the source of the plant name that would have been borrowed.

There are over 100 ethnic groups and languages in Nigeria, and each has different language names for Yam, "*Isu*" is the Yoruba translation or "*Iyan*" when it has been prepared to be consumed as a main course for dinner.

Other Uses of the Term Yam

Several other unrelated root vegetables are sometimes referred to as "yams", including:

- In the United States, sweet potato (*Ipomoea batatas*), especially those with orange flesh, are often referred to as "yams". In the United States, firm varieties of sweet potatoes were produced before soft varieties. When soft varieties were first grown commercially, there was a need to differentiate between the two. African slaves had already been calling the soft sweet potatoes "yams" because they resembled the yams in Africa. Thus, soft sweet potatoes were referred to as yams to distinguish them from the firm varieties. Sweet potatoes labelled as "yams" are widely available in markets that serve Asian or Caribbean communities. Today the US Department of Agriculture requires sweet potatoes labelled with the term "yam" to be accompanied by the term "sweet potato".

- Just like commercial yams in the United States, Okinawan purple "yam" is actually a sweet potato.
- In New Zealand "yam" refers to the oca (*Oxalis tuberosa*), while in the Solomon Islands and other Melanesian countries the yam is also known as the "common desert truffle".
- The corm of the konjac is often colloquially referred to as a yam, although it is unrelated.
- In Malaysia and Singapore, "yam" is used to refer to taro.

Euphorbiaceae

Euphorbiaceae, the Spurge family are a large family of flowering plants with 300 genera and around 7,500 species. Most are herbs, but some, especially in the tropics, are also shrubs or trees. Some are succulent and resemble cacti.

This family occurs mainly in the tropics, with the majority of the species in the Indo-Malayan region and tropical America a good second. There is a large variety in tropical Africa, but it is not as abundant or varied as in these two other tropical regions. However, *Euphorbia* also has many species in non-tropical areas such as the Mediterranean Basin, the Middle East, South Africa, and southern USA.

The leaves are alternate, seldom opposite, with stipules. They are mainly simple, but where compound, are always palmate, never pinnate. Stipules may be reduced to hairs, glands, or spines, or in succulent species are sometimes absent.

The radially symmetrical flowers are unisexual, with the male and the female flowers usually occurring on the same plant. As can be expected from such a large family, there is a wide variety in the structure of the flowers. They can be monoecious or dioecious. The stamens (the male organs) can number from 1 to 10 (or even more). The female flowers are hypogynous, that is, with a superior ovary.

The genera in tribe Euphorbieae, subtribe Euphorbiinae (*Euphorbia* and close relatives) show a highly specialised form of pseudanthium ("false flower" made up of several true flowers) called a cyathium. This is usually a small cup-like involucre consisting of fused together bracts and peripheral nectary glands, surrounding a ring of male flowers, each a single stamen. In the middle of the cyathium stands a female flower: a single pistil with branched stigmas. This whole arrangement resembles a single flower.

The fruit is usually a schizocarp, sometimes a drupe. A typical schizocarp is the regma, a capsular fruit with three or more cells, each of which splits open at maturity into separate parts and then breaks away explosively, scattering the small seeds. The family contains a large variety of phytotoxins (toxic substances produced by plants), mainly diterpene esters, alkaloids, glycosides, and ricin-type toxins. A milky sap or latex is a characteristic of the subfamilies Euphorbioideae and Crotonoideae. This milky sap is poisonous in the Euphorbioideae, but innocuous in the Crotonoideae. White mangrove (*Excoecaria agallocha*), or Blind-Your-Eye Mangrove sap causes blistering on contact and temporary blindness if it contacts the eyes. Other common names are Milky Mangrove, Buta Buta (Malay), Gewa (Bangladesh). The sap of spurge was used as a laxative.

Recent molecular studies have shown that the enigmatic family Rafflesiaceae, which was only recently recognised to belong to order Malpighiales, is derived from within Euphorbiaceae (Davis et al. 2007).

Uses

A number of plants of the Spurge family are of considerable economic importance. Prominent plants include Cassava (*Manihot esculenta*), Castor oil plant (*Ricinus communis*), Barbados nut (*Jatropha curcas*), and the Para rubber tree (*Hevea brasiliensis*). Many are grown as ornamental plants, such as Poinsettia (*Euphorbia pulcherrima*). Leafy spurge (*Euphorbia esula*) and Chinese tallow (*Sapium sebiferum*) are invasive weeds in North America. In medicine, some species of Euphorbiaceae proved effective against genital herpes (HSV-2).

Vulnerability

Some species, despite their medicinal benefits, are facing the risk of becoming extinct. These species include the *Euphorbia* species *E. appariciana*, *E. attastoma*, *E. crossadenia* and *E. gymnoclada*.

3

Varieties of Vegetables

Common Onion Group (Var. *Cepa*)

Most of the diversity within *A. cepa* occurs within this group, the most economically important *Allium* crop. Plants within this group form large single bulbs, and are grown from seed or seed-grown sets. The majority of cultivars grown for dry bulbs, salad onions, and pickling onions belong to this group. The range of diversity found among these cultivars includes variation in photoperiod (length of day that triggers bulbing), storage life, flavour, and skin colour. Common onions range from the pungent varieties used for dried soups and onion powder to the mild and hearty sweet onions, such as the Vidalia from Georgia, USA, or Walla Walla from Washington that can be sliced and eaten on a sandwich instead of meat.

European Onions

A number of onions have Protected Geographical Status in Europe, these include:

- Cebolla Fuentes de Ebro, a sweet, watery and soft variety from Zaragoza, Spain (*PDO*)
- Cipolla Rossa di Tropea, a red onion from Calabria, Italy (*PGI*)
- Cipollotto Nocerino, a spring/salad onion-sized *Allium cepa* from Campania, Italy (*PDO*)
- Oignon doux des Cévennes, a sweet onion from the southeast of France (*PDO*)

Aggregatum Group (Var. Aggregatum)

This group contains shallots and potato onions, also referred to as multiplier onions. The bulbs are smaller than those of common onions, and a single plant forms an aggregate cluster of several bulbs. They are propagated almost exclusively from daughter bulbs, although reproduction from seed is possible. Shallots are the most important subgroup within this group and comprise the only cultivars cultivated commercially. They form aggregate clusters of small, narrowly ovoid to pear-shaped bulbs. Potato onions differ from shallots in forming larger bulbs with fewer bulbs per cluster, and having a flattened (onion-like) shape. However, intermediate forms exist. In Nepali onion is called 'pyaj'.

Species that may be Confused with A. Cepa

Scallions or salad onions may be grown from the Welsh onion (*A. fistulosum*) as well as from *A. cepa*. Young plants of *A. fistulosum* and *A. cepa* look very similar, but may be distinguished by their leaves, which are circular in cross-section in *A. fistulosum* rather than flattened on one side.

Hybrids with A. Cepa Parentage

A number of hybrids are cultivated that have A. cepa parentage, such as the tree onion or Egyptian onion (A. ×proliferum), Wakegi onion (A. ×wakegi), and the triploid onion (A. ×cornutum). The tree onion or Egyptian onion produces bulblets in the flower head instead of flowers, and is now known to be a hybrid of A. cepa × A. fistulosum. It has previously been treated as a variety of A. cepa, for example A. cepa var. proliferum, A. cepa var. bulbiferum, and A. cepa var. viviparum.

The Wakegi onion is also known to be a hybrid between A. cepa and A. fistulosum, with the A. cepa parent believed to be from the Aggregatum Group of cultivars. It has been grown for centuries in Japan and China for use as a salad onion. Under the rules of botanical nomenclature, both the Egyptian onion and Wakegi onion should be combined into one hybrid species, having the same parent species. Where this is followed, the Egyptian onion is named A. ×proliferum Eurasian Group and the Wakegi onion is named A. ×proliferum East Asian Group.

Triploid Onions

The triploid onion is a hybrid species with three sets of chromosomes, two sets from *A. cepa* and the third set from an unknown parent. Various clones of the triploid onion are grown locally in different regions, such as 'Ljutika' in Croatia, and 'Pran', 'Poonch' and 'Srinagar' in the India-Kashmir region. 'Pran' is grown extensively in the Northern

Indian provinces of Jammu and Kashmir. There are very small genetic differences between 'Pran' and the Croatian clone 'Ljutika', implying a monophyletic origin for this species. Some authors have used the name *A. cepa* var. *viviparum* (Metzg.) Alef. for the triploid onion, but this name has also been applied to the Egyptian onion. The only name unambiguously connected with the triploid onion is *A. ×cornutum*.

Storage

Green onion and leeks are optimally stored refrigerated. Cooking onions and sweet onions, on the other hand, can be stored at room temperature, optimally in a single layer, in mesh bag in a dry, cool, dark, well ventilated location. In this environment, cooking onions have a shelf life of three to four weeks, and sweet onions one to two weeks.

Cooking onions will absorb odours from apples and pears. Also, they draw moisture from vegetables they are stored with which may cause them to decay. Sweet onions have a greater water and sugar content than cooking onions. This makes them sweeter and milder tasting, but also reduces their shelf life. Sweet onions can also be stored refrigerated; they have a shelf life of approximately one month, optimally uncovered. Irrespective of type, any cut pieces of onion are optimally tightly wrapped, stored away from other produce, and used within two to three days.

Production Trends

The Onion Futures Act, passed in 1958, bans the trading of futures contracts on onions in the United States, after farmers complained about alleged market manipulation by Sam Seigel and Vincent Kosuga at the Chicago Mercantile Exchange. It provides economists with a unique case study in the effects of futures trading on agricultural prices. It remains in effect as of 2012.

Amaranthaceae

The Amaranthaceae, the Amaranth family, represent the most species-rich lineage within the flowering plant order of Caryophyllales. Now including the former goosefoot family (Chenopodiaceae), the extended family contains approximately 180 genera and 2,500 species.

Description

Vegetative Characters

Most of these species are annual or perennial herbs or subshrubs, some are shrubs; very few species are vines or trees. Some species are succulent. Many species have stems with thickened nodes. The wood of the perennial stem has a typical "anomalous" secondary growth, only in subfamily Polycnemoideae there is normal secondary growth. The leaves are mostly alternate, sometimes opposite. They never possess stipules. The simple leaves are flat or terete, their shape is extremely variable, with entire or toothed margins. In some species, the leaves are reduced to minute scales. In most cases, neither basal or terminal aggregations of leaves occur.

Inflorescence and Flowers

The flowers are solitary or aggregated in cymes, spikes, or panicles and typically perfect (bisexual) and actinomorphic. Some species have unisexual flowers. Bracts and bracteoles are either herbaceous or scarious. Flowers are regular with a herbaceous or scarious perianth of (1 to) mostly 5 (rarely to 8) tepals, often joined.

There are 1-5 stamens opposite to tepals or alternating, inserting from a hypogynous disc, which may have appendages (pseudostaminodes) in some species. The anthers have 2 or 4 pollen sacs (locules). In tribe *Caroxyloneae*, antheres have vesicular appendages. The pollen grains are sperical with many pores (pantoporate), with pore numbers from a few to 250 (in *Froelichia*). 1-3 (rarely -6) carpels are fused to a superior ovary with 1 (rarely 2) basal ovule.

Fruits and Seeds

The diaspores are seeds ore fruits (utricles), more often the perianth persists and is modified in fruit for means of dispersal. Sometimes even bracts and bracteoles may belong to the diaspore. More rarely the fruit is a circumscissile capsule or a berry. The horizontal or vertical seed often has a thickened or woody seed coat. The green or white embryo is either spirally (and without perisperm) or annular (rarely straight).

Phytochemistry

Widespread in the Amaranthaceae is the occurrence of betalain pigments. The former Chenopodiaceae often contain isoflavonoids. In phytochemical research, several methylenedioxyflavonols, saponins, triterpenoids, ecdysteroids, and specific rootlocated carbohydrates have been found.

Photosynthesis Pathway

With approximately 800 species that are C_4-plants, the Amaranthaceae represent the largest group with this photosynthesis pathway. Within the family, there are several types of C_4-photosynthesis, and about 17 different types of leaf anatomy are realised. Therefore this photosynthesis pathway seems to have been developed about 15 times independently during the evolution of the family. Two-third of the C_4-species belong to the former Chenopodiaceae. The first occurrence of C_4-photosynthesis dates from the early miocene, about 24 million years ago. But in some groups this photosynthesis pathway has evolved much later, about 6 (or less) million years ago.

The multiple origin of C_4-photosynthesis in Amaranthaceae is regarded as an evolutionary response to a permanent shortage in water supply in combination with high temperatures. These species with their higher water use efficiency had a selective advantage and were able to spread out into dry (arid) habitats.

Distribution

This is a widespread and cosmopolitan family from the tropics to cool temperate regions. The Amaranthaceae (sensu strictu) are predominantly tropical, whereas the former Chenopodiaceae have their centres of diversity in dry temperate and warm temperate areas. Many of the species are halophytes, tolerating salty soils, or grow in dry steppes or semideserts.

Economic Importance

Some species, like spinach (*Spinacia oleracea*) or forms of beet (*Beta vulgaris*) (Beetroot, Chard), are used as vegetables. Forms of *Beta vulgaris* are fodder beet (Mangelwurzel) and Sugar beet. The seeds of *Amaranthus* and Quinoa (*Chenopodium quinoa*) are edible and used like cereals. *Dysphania ambrosioides* and *Dysphania anthelmintica* are used as medicinal herbs. Several species are used as a source of soda ash. A number of species are popular garden ornamental plants, especially species from *Alternanthera*, *Amaranthus*, *Celosia*, and *Iresine*. Some species are considered weeds.

Systematics

In the APG II system, of 2003 (unchanged from the APG system, of 1998), the family is placed in the order Caryophyllales. It includes the plants formerly treated as the family Chenopodiaceae. The monophyly of this new, broadly defined Amaranthaceae has been strongly supported by both morphological and phylogenetic analyses.

The family Amaranthaceae was first published in 1789 by Antoine Laurent de Jussieu in *Genera Plantarum*, p. 87–88. The first publication of family Chenopodiaceae was in 1799 by Étienne Pierre Ventenat in *Tableau du Regne Vegetal*, 2, p. 253. The older name has priority and is now the valid scientific name of the extended Amaranthaceae (s.l. = sensu latu).

Many recent publications are still referring to the family name Chenopodiaceae. Phylogenetic research revealed the important impact of the subfamily Polycnemoideae on the classification: If Polycnemoideae are considered being part of Chenopodiaceae, then Amaranthaceae (s.str.=sensu strictu) have to be included, too, and the name of the extended family is Amaranthaceae. If Polycnemoideae would be separated as an own family, Chenopodiaceae and Amaranthaceae (s.str.) would form two distinct monophyletic groups and could be treated as two separate families.

Amaranthaceae Juss. (s.l.) include the former families Achyranthaceae Raf., Atriplicaceae Durande, Betaceae Burnett, Blitaceae T.Post & Kuntze, Celosiaceae Martynov, Chenopodiaceae Vent., nom. cons., Corispermaceae Link, Deeringiaceae J.Agardh, Dysphaniaceae (Pax) Pax, nom. cons., Gomphrenaceae Raf., Polycnemaceae Menge, Salicorniaceae Martynov, Salsolaceae Menge, and Spinaciaceae Menge.

The systematics of Amaranthaceae is subject of intensive recent research. Molecular genetic studies revealed, that the traditional classification, based on morphological and anatomical characters, often did not reflect the phylogenetical relationships.

The former Amaranthaceae (in their narrow circumscription) are classified in two subspecies, Amaranthoideae and Gomphrenoideae, and contain about 65 genera and 900 species in tropical Africa and North America. The Amaranthoideae and some genera of Gomphrenoideae were found to be polyphyletic, so taxonomical changes are needed.

Current studies classified the species of former Chenopodiaceae to 8 distinct subfamilies (the research is not yet completed): Polycnemoideae, which are regarded as a basal lineage, Betoideae, Camphorosmoideae, Chenopodioideae, Corispermoideae, Salicornioideae, Salsoloideae, and Suaedoideae. In this preliminary classification, the Amaranthaceae s.l. are divided into 10 subfamilies with approximately 180 genera and 2,500 species.

Amaranth

Amaranthus, collectively known as amaranth, is a cosmopolitan genus of herbs. Approximately 60 species are recognised, with inflorescences and foliage ranging from purple and red to gold. Members of this genus share many characteristics and uses with members of the closely related genus *Celosia*.

Although several species are often considered weeds, people around the world value amaranths as leaf vegetables, cereals, and ornamentals. It is also known as "love lies bleeding" — it was used in the Middle Ages to stem bleeding. The ultimate root of "amaranth" is the (*amarantos*), "unfading," with the "flower," (*anthos*), factoring into the word's development as "amaranth." The more accurate "amarant" is an archaic variant.

Systematics

Amaranthus shows a wide variety of morphological diversity among and even within certain species. Although the family (Amaranthaceae) is distinctive, the genus has few distinguishing characters among the 70 species included. This complicates taxonomy and *Amaranthus* has generally been considered among systematists as a "difficult" genus.

Formerly, Sauer (1955) classified the genus into two subgenera, differentiating only between monoecious and dioecious species: *Acnida* (L.) Aellen ex K.R. Robertson and *Amaranthus*. Although this classification was widely accepted, further infrageneric classification was (and still is) needed to differentiate this widely diverse group.

Currently, *Amaranthus* includes three recognised subgenera and 70 species, although species numbers are questionable due to hybridization and species concepts. Infrageneric classification focuses on inflorescence, flower characters and whether a species is monoecious/dioecious, as in the Sauer (1955) suggested classification. A modified infrageneric classification of *Amaranthus* was published by Mosyakin & Robertson (1996) and includes three subgenera: *Acnida*, *Amaranthus*, and *Albersia*. The taxonomy is further differentiated by sections within each of the subgenera.

Species

- *Amaranthus acanthochiton* – greenstripe
- *Amaranthus acutilobus* – sharp-lobe amaranth; is a synonym of *Amaranthus viridis*
- *Amaranthus albus* – white pigweed, prostrate pigweed, pigweed amaranth
- *Amaranthus arenicola* – sandhill amaranth
- *Amaranthus australis* – southern amaranth

- *Amaranthus bigelovii* – Bigelow's amaranth
- *Amaranthus blitoides* – mat amaranth, prostrate amaranth, prostrate pigweed
- *Amaranthus blitum* – purple amaranth
- *Amaranthus brownii* – Brown's amaranth
- *Amaranthus californicus* – California amaranth, California pigweed
- *Amaranthus cannabinus* – tidal-marsh amaranth
- *Amaranthus caudatus* – love-lies-bleeding, pendant amaranth, tassel flower, *quilete*
- *Amaranthus chihuahuensis* – Chihuahuan amaranth
- *Amaranthus chlorostachys*
- *Amaranthus crassipes* – spreading amaranth
- *Amaranthus crispus* – crispleaf amaranth
- *Amaranthus cruentus* – purple amaranth, red amaranth, Mexican grain amaranth
- *Amaranthus deflexus* – large-fruit amaranth
- *Amaranthus dubius* – spleen amaranth, *khada sag*
- *Amaranthus fimbriatus* – fringed amaranth, fringed pigweed
- *Amaranthus floridanus* – Florida amaranth
- *Amaranthus gangeticus* – elephant head amaranth
- *Amaranthus graecizans*
- *Amaranthus greggii* – Gregg's amaranth
- *Amaranthus hybridus* – smooth amaranth, smooth pigweed, red amaranth
- *Amaranthus hypochondriacus* – Prince-of-Wales feather, prince's feather
- *Amaranthus leucocarpus*
- *Amaranthus lineatus* – Australian amaranth
- *Amaranthus lividus*
- *Amaranthus mantegazzianus* – *quinoa de Castilla*
- *Amaranthus minimus*
- *Amaranthus muricatus* – African amaranth
- *Amaranthus obcordatus* – Trans-Pecos amaranth
- *Amaranthus oleraceous* – *kosala sag*

- *Amaranthus palmeri* – Palmer's amaranth, Palmer pigweed, careless weed
- *Amaranthus paniculus* – Reuzen amarant
- *Amaranthus polygonoides* – tropical amaranth
- *Amaranthus powellii* – green amaranth, Powell amaranth, Powell pigweed
- *Amaranthus pringlei* – Pringle's amaranth
- *Amaranthus pumilus* – seaside amaranth
- *Amaranthus quitensis* – *ataco, sangorache*
- *Amaranthus retroflexus* – red-root amaranth, redroot pigweed, common amaranth
- *Amaranthus rudis* – tall amaranth, common waterhemp
- *Amaranthus scleropoides* – bone-bract amaranth
- *Amaranthus spinosus* – spiny amaranth, prickly amaranth, thorny amaranth
- *Amaranthus standleyanus*
- *Amaranthus thunbergii* – Thunberg's amaranth
- *Amaranthus torreyi* – Torrey's amaranth
- *Amaranthus tricolour* – Joseph's-coat
- *Amaranthus tuberculatus* – rough-fruit amaranth, tall waterhemp
- *Amaranthus viridis* – slender amaranth, green amaranth
- *Amaranthus watsonii* – Watson's amaranth
- *Amaranthus wrightii* – Wright's amaranth

Uses

Amaranth Seed: Several species are raised for amaranth "grain" in Asia and the Americas. This should more correctly be termed "pseudograin". Ancient amaranth grains still used to this day include the three species, *Amaranthus caudatus*, *Amaranthus cruentus*, and *Amaranthus hypochondriacus*. Although amaranth was cultivated on a large scale in ancient Mexico, Guatemala, and Peru, nowadays it is only cultivated on a small scale there, along with India, China, Nepal, and other tropical countries; thus, there is potential for further cultivation in those countries, as well as in the US.

In a 1977 article in *Science*, amaranth was described as "the crop of the future". It has been proposed as an inexpensive native crop that could be cultivated by indigenous people in rural areas for several reasons:

1. It is easily harvested.
2. Its seeds are a good source of protein. Compared to other grains, amaranth is unusually rich in the essential amino acid lysine Common grains such as wheat and corn are comparatively rich in amino acids that amaranth lacks; thus, amaranth and other grains can complement each other.
3. The seeds of *Amaranthus* species contain about thirty percent more protein than cereals like rice, sorghum and rye. In cooked and edible forms, amaranth is competitive with wheat germ and oats - higher in some nutrients, lower in others.
4. It is easy to cook. As befits its weedy life history, amaranth grains grow very rapidly and their large seedheads can weigh up to 1 kilogram and contain a half-million seeds in three species of amaranth.

Kiwicha, as amaranth is known today in the Andes, was one of the staple foodstuffs of the Incas. Known to the Aztecs as huautli, it is thought to have represented up to 80% of their caloric consumption before the conquest. Another important use of amaranth throughout Mesoamerica was to prepare ritual drinks and foods.

To this day, amaranth grains are toasted much like popcorn and mixed with honey, molasses or chocolate to make a treat called *alegría*, meaning "joy" in Spanish.

Diego Duran described the festivities for Huitzilopochtli, a blue hummingbird god. (Real hummingbirds feed on amaranth flowers.)

The Aztec month of Panquetzaliztli (7 December to 26 December) was dedicated to Huitzilopochtli. People decorated their homes and trees with paper flags; there were ritual races, processions, dances, songs, prayers, and finally human sacrifices.

This was one of the more important Aztec festivals, and the people prepared for the whole month.

They fasted or ate very little; a statue of the god was made out of amaranth (*huautli*) seeds and honey, and at the end of the month, it was cut into small pieces so everybody could eat a little piece of the god. After the Spanish conquest, cultivation of amaranth was outlawed, while some of the festivities were subsumed into the Christmas celebration. Because of its importance as a symbol of indigenous culture, its gluten-free palatability, easy to cook, and its protein particularly well suited to human nutritional needs, interest in grain amaranth (especially *A. cruentus* and *A. hypochondriacus*) revived in the 1970s.

It was recovered in Mexico from wild varieties and is now commercially cultivated. It is a popular snack sold in Mexico City and other parts of Mexico, sometimes mixed with chocolate or puffed rice, and its use has spread to Europe and parts of North America. Amaranth and quinoa are called pseudograins because of their flavour and cooking similarities to grains.

Amaranth Seed Flour: Amaranth seed flour has been evaluated as an additive to wheat flour by food specialists. To determine palatability, different levels of amaranth grain flour were mixed with the wheat flour and baking ingredients (1% salt, 2.5% fat, 1.5% yeast, 10% sugar and 52–74% water), fermented, molded, pan-proved and baked. The baked products were evaluated for loaf volume, moisture content, colour, odor, taste and texture.

The amaranth containing products were then compared with bread made from 100% wheat flour. The loaf volume decreased by 40% and the moisture content increased from 22 to 42% with increase in amaranth grain flour. The study found that the sensory scores of the taste, odor colour and texture decreased with increasing amounts of amaranth. Generally, above 15% amaranth grain flour, there were significant differences in the evaluated sensory qualities and the high amaranth-containing product was found to be of unacceptable palatability to the population sample that evaluated the baked products.

Leaves, Roots, and Stems

Amaranth species are cultivated and consumed as a leaf vegetable in many parts of the world. There are four species of *Amaranthus* documented as cultivated vegetables in eastern Asia: *Amaranthus cruentus, Amaranthus blitum, Amaranthus dubius*, and *Amaranthus tricolour*.

In Indonesia and Malaysia, leaf amaranth is called *bayam*, while the Tagalogs in the Philippines call the plant *kilitis* or “kulitis”. The Ilocanos (Philippines) call it “kalunay”. In the state of Uttar Pradesh and Bihar in India, it is called Chaulai and is a popular green leafy vegetable (referred to in the class of vegetable preparations called Saag). It is called Chua in Kumaun area of Uttarakhand, where it is a popular red-green vegetable. In Karnataka state in India, it is called Harive (ಹರಿವೆ). It is used to prepare curries like Hulee, palya, Majjigay-hulee and so on. In the state of Kerala, it is called ‘Cheera’ and is consumed by stir-frying the leaves with spices and red chillies to make 'Cheera Thoran'. In TamilNadu State, it is called முளைக்கீரை and is regularly consumed as a favourite dish, where the greens are steamed, and mashed, with light seasoning of salt, red chillis and cumin. It is called *keerai masial* (கீரை மசியல்). In Andhra Pradesh, India, this leaf is added in preparation of a popular dal called *thotakura pappu* తోట కూర పప్పు. In Maharashtra, it is called as "Shravani Maath" and it is available in both red and white colour. In Orissa, it is called as "Khada saga", it is used to prepare 'Saga Bhaja', in which the leaf is fried with chillies and onions.

The root of mature amaranth is a popular vegetable. It is white and cooked with tomatoes or tamarind gravy. It has a milky taste and is alkaline.

In China, the leaves and stems are used as a stir-fry vegetable, or in soups, and called 苋菜. Amaranth greens are believed to help enhance eyesight. In Vietnam, it is called *rau dền* and is used to make soup. There are two species popular as edible vegetable in Vietnam: *dền đỏ*-amaranthus tricolour and *dền cõm* or *dền trắng*- amaranthus viridis.

A traditional food plant in Africa, amaranth has the potential to improve nutrition, boost food security, foster rural development and support sustainable land care. In East Africa, Amaranth leaf is known in chewa as *bonongwe,* and in Swahili as *mchicha,* as *terere* in Kikuyu, Meru and Embu; and as *telele* in Kamba. In Bantu regions of Uganda it is known as *doodo*. It is recommended by some doctors for people having low red blood cell count. It is also known among the Kalenjin as a drought crop (*chepkerta*). In Lingala (spoken in the Democratic Republic of the Congo and the Republic of Congo), it is known as *l[ngal[nga* or *bít[kut[ku*. In Nigeria, it is a common vegetable and goes with all Nigerian starch dishes. It is known in Yoruba as *efo tete* or *arowo jeja* (meaning "we have money left over for fish"). In the Caribbean, the leaves are called *bhaji* in Trinidad and *callaloo* in Jamaica and are stewed with onions, garlic and tomatoes, or sometimes used in a soup called pepperpot soup.

In Greece, green amaranth (*Amaranthus viridis*) is a popular dish and is called *vlita* or *vleeta*. It is boiled, then served with olive oil and lemon like a salad, usually alongside fried fish. Greeks stop harvesting the plant (which usually grows wild) when it starts to bloom at the end of August. In Sri Lanka, it is called "koora thampala". Sri Lankans cook it and eat it with rice. Fiji Indians call it *choraiya bhaji*.

Dyes

The flowers of the 'Hopi Red Dye' amaranth were used by the Hopi (a tribe in the western United States) as the source of a deep red dye. There is also a synthetic dye that has been named "amaranth" for its similarity in colour to the natural amaranth pigments known as betalains. This synthetic dye is also known as Red No. 2 in North America and E123 in the European Union.

Ornamentals

The genus also contains several well-known ornamental plants, such as *Amaranthus caudatus* (love-lies-bleeding), a native of India and a vigorous, hardy annual with dark purplish flowers crowded in handsome drooping spikes. Another Indian annual, *A. hypochondriacus* (prince's feather), has deeply veined lance-shaped leaves, purple on the under face, and deep crimson flowers densely packed on erect spikes.

Amaranths are recorded as food plants for some Lepidoptera (butterfly and moth) species including the nutmeg moth and various case-bearer moths of the genus *Coleophora*: *C. amaranthella*, *C. enchorda* (feeds exclusively on *Amaranthus*), *C. immortalis* (feeds exclusively on *Amaranthus*), *C. lineapulvella* and *C. versurella* (recorded on *A. spinosus*).

Nutritional Value

Amaranth greens, also called Chau lai (Hindi) and Chu or Chua (Kumauni), Chinese spinach, hinn choy or yin tsoi callaloo, dhantinasoppu / harive (Kannada); (Telugu); Rajgira (Tamil); cheera (Malayalam); bayam (Indonesian); phak khom (Thai); tampala, or quelite (Oriya); Khada Saga, are a common leaf vegetable throughout the tropics and in many warm temperate regions.

Cooked amaranth leaves are a good source of vitamin A, vitamin C, and folate; they are also a complementing source of other vitamins such as thiamine, niacin, and riboflavin, plus some dietary minerals including calcium, iron, potassium, zinc, copper, and manganese.

Cooked amaranth grains are a complementing source of thiamine, niacin, riboflavin, and folate, and dietary minerals including calcium, iron, magnesium, phosphorus, zinc, copper, and manganese - comparable to common grains such as wheat germ, oats and others. Amaranth seeds contain lysine, an essential amino acid, limited in other grains or plant sources.

Most fruits and vegetables do not contain a complete set of amino acids, and thus different sources of protein must be used. Amaranth too is limited in some essential amino acids, such as leucine and threonine. Amaranth seeds are therefore promising complement to common grains such as wheat germ, oats, corn because these common grains are abundant sources of essential amino acids found to be limited in amaranth. Amaranth may be a promising source of protein to those who are gluten sensitive, because unlike the protein found in grains such as wheat and rye, its protein does not contain gluten. According to a 2007 report, amaranth compares well in nutrient content with gluten-free vegetarian options such as buckwheat, corn, millet, wild rice, oats and quinoa.

Several studies have shown that like oats, amaranth seed or oil may be of benefit for those with hypertension and cardiovascular disease; regular consumption reduces blood pressure and cholesterol levels, while improving antioxidant status and some immune parameters. While the active ingredient in oats appears to be water-soluble fiber, amaranth appears to lower cholesterol via its content of plant stanols and squalene.

Amaranth remains an active area of scientific research for both human nutritional needs and foraging applications. Over 100 scientific studies suggest a somewhat conflicting picture on possible anti-nutritional and toxic factors in amaranth, more so in some particular strains of amaranth.

Lehmann, in a review article, identifies some of these reported anti-nutritional factors in amaranth to be phenolics, saponins, tannins, phytic acid, oxalates, protease inhibitors, nitrates, polyphenols and phytohemagglutinins. Of these, oxalates and nitrates are of more concern when amaranth grain is used in foraging applications.

Some studies suggest thermal processing of amaranth, particularly in moist environment, prior to its preparation in food and human consumption may be a promising way to reduce the adverse effects of amaranth's anti-nutritional and toxic factors.

As a Weed

Not all amaranth plants are cultivated. Most of the species from *Amaranthus* are summer annual weeds and are commonly referred to as pigweeds. These species have an extended period of germination, rapid growth, and high rates of seed production, and have been causing problems for farmers since the mid-1990s.

This is partially due to the reduction in tillage, reduction in herbicidal use and the evolution of herbicidal resistance in several species where herbicides have been applied more often. The following

9 species of *Amaranthus* are considered invasive and noxious weeds in the US and Canada: *A. albus*, *A. blitoides*, *A. hybridus*, *A. palmeri*, *A. powellii*, *A. retroflexus*, *A. spinosus*, *A. tuberculatus*, and *A. viridis*.

A new herbicide-resistant strain of Amaranthus palmeri or Palmer amaranth has appeared; it is Glyphosate-resistant and so cannot be killed by the widely used Roundup herbicide. Also, this plant can survive in tough conditions. This could be of particular concern to cotton farmers using Roundup Ready cotton. The species *Amaranthus palmeri* (Palmer amaranth) causes the greatest reduction in soybean yields and has the potential to reduce yields by 17-68% in field experiments.

Palmer amaranth is among the "top five most troublesome weeds" in the southeast of the United States and has already evolved resistances to dinitroanilines and acetolactate synthase inhibitors. This makes the proper identification of *Amaranthus* species at the seedling stage essential for agriculturalists. Proper weed control needs to be applied before the species successfully colonizes in the crop field and causes significant yield reductions.

Beneficial Weed

Pigweed can be a beneficial weed, as well as a companion plant, serving as a trap for leaf miners and some other pests, as well as sheltering ground beetles (which prey upon insect pests) and breaking up hard soil for more delicate neighbouring plants.

Taxonomy

Although widely recognised in most plant classifications (notably the Cronquist system), the APG system (1998) and the APG II system (2003) have included these plants in the family Amaranthaceae on the basis of evidence from molecular phylogenies: they are a paraphyletic group, so that for a monophyletic taxon, merger with Amaranthaceae *sensu stricto* was needed.

Current studies have classified the species of Chenopodiaceae to several distinct subfamilies (Betoideae, Camphorosmoideae, Chenopodioideae, Polycnemoideae, Salicornioideae, Salsoloideae and Suaedoideae).

Pollen Allergy

Chenopod pollen is a common allergen. Most species bloom in summer months. Pollen at the microscopic level is mostly indistinguishable from species to species, but cross-reactivity has not been definitively established. Because flowering does not occur in the commercial planting cycle, localized pollen counts will not increase due to planting of Chenopod crops.

Basella Alba

Basella alba, or Malabar spinach, Malabar nightshade, Alugbati or Alabati in Philippines, (also Phooi leaf, Red vine spinach, Creeping spinach, Climbing spinach, Indian spinach, Philippine Spinach, Asian Spinach) is a perennial vine found in the tropics where it is widely used as a leaf vegetable.

Description

Basella alba is a fast-growing, soft-stemmed vine, reaching 10 m in length. Its thick, semi-succulent, heart-shaped leaves have a mild flavour and mucilaginous texture. The stem of the cultivar *Basella alba* 'Rubra' is reddish-purple.

Soil and Climate Requirements

Basella alba grows well under full sunlight in hot, humid climates and in areas lower than 500 m above sea level. Growth is slow in low temperatures resulting in low yields. Flowering is induced during the short-day months of November to February. It grows best in sandy loam soils rich in organic matter with pH ranging from 5.5 to 8.0.

Food Uses

Typical of leaf vegetables, Malabar spinach is high in vitamin A, vitamin C, iron, and calcium. It is low in calories by volume, but high in protein per calorie. The succulent mucilage is a particularly rich source of soluble fiber. Among many other possibilities, Malabar spinach may be used to thicken soups or stir-fries with garlic and chili peppers. In Bengali cuisine it is widely used to cook with Ilish fish.

In Andhra Pradesh, a southern state in India, a curry of Basella and Yam is made popularly known as Kanda Bachali Koora (Yam and Basella curry). The vegetable is used in Chinese cuisine.

Its many names include flowing water vegetable. In Vietnam, particularly the north, it is cooked with crab meat, luffa and jute to make soup. In Orissa, India, it is used to make Curries and Saaga (any type of dish made from green leafy vegetables is called Saaga in Orissa). In Maharashtra, India, it is used to make bhaji. In Africa, the mucilaginous cooked shoots are most commonly used. Malabar spinach can be found at many Chinese/Vietnamese/Korean/Indian grocery stores, as well as farmers' markets. It has been shown to contain certain phenolic phytochemicals and it has antioxidant properties.

Sweet Potato

The sweet potato (*Ipomoea batatas*) is a dicotyledonous plant that belongs to the family Convolvulaceae. Its large, starchy, sweet-tasting, tuberous roots are an important root vegetable.

The young leaves and shoots are sometimes eaten as greens. Of the approximately 50 genera and more than 1,000 species of Convolvulaceae, *I. batatas* is the only crop plant of major importance—some others are used locally, but many are actually poisonous. The sweet potato is only distantly related to the potato (*Solanum tuberosum*).

Although the soft, orange sweet potato is often mislabelled a yam in parts of North America, the sweet potato is botanically very distinct from a genuine yam, which is native to Africa and Asia and belongs to the monocot family Dioscoreaceae. To prevent confusion, the United States Department of Agriculture requires sweet potatoes labelled as "yams" to be labelled also as "sweet potatoes". The genus *Ipomoea* that contains the sweet potato also includes several garden flowers called morning glories, though that term is not usually extended to *Ipomoea batatas*. Some cultivars of *Ipomoea batatas* are grown as ornamental plants; the name "tuberous morning glory" may be used in a horticultural context.

The plant is a herbaceous perennial vine, bearing alternate heart-shaped or palmately lobed leaves and medium-sized sympetalous flowers. The edible tuberous root is long and tapered, with a smooth skin whose colour ranges between yellow, orange, red, brown, purple, and beige. Its flesh ranges from beige through white, red, pink, violet, yellow, orange, and purple. Sweet potato varieties with white or pale yellow flesh are less sweet and moist than those with red, pink or orange flesh. In certain parts of the world, sweet potatoes are locally known as camote, kamote, goguma, man thet, ubi jalar, ubi keladi, shakarkand, satsuma imo, batata or el boniato. In New Zealand English, the Mâori term *kûmara* is commonly used.

Origin, Distribution and Diversity

The centre of origin and domestication of sweet potato is thought to be either in Central America or South America. In Central America, sweet potatoes were domesticated at least 5,000 years ago. In South

America, Peruvian sweet potato remnants dating as far back as 8000 BCE have been found.

Austin (1988) postulated that the centre of origin of *I. batatas* was between the Yucatán Peninsula of Mexico and the mouth of the Orinoco River in Venezuela. The 'cultigen' had most likely been spread by local people to the Caribbean and South America by 2500 BCE. Zhang *et al.* (1998) provided strong supporting evidence that the geographical zone postulated by Austin is the primary centre of diversity. The much lower molecular diversity found in Peru–Ecuador suggests this region should be considered as secondary centre of sweet potato diversity.

The sweet potato was also grown before western exploration in Polynesia. Sweet potato has been radiocarbon-dated in the Cook Islands to 1000 CE, and current thinking is that it was brought to central Polynesia around 700 CE, possibly by Polynesians who had travelled to South America and back, and spread across Polynesia to Hawaii and New Zealand from there.

It is possible, however, that South Americans brought it to the Pacific, although this is unlikely as it was the Polynesians who had a strong maritime tradition and not the native South Americans. The theory that the plant could spread by floating seeds across the ocean is not supported by evidence. Another point is that the sweet potato in Polynesia is the cultivated *Ipomoea batatas*, which is generally spread by vine cuttings and not by seeds. Sweet potatoes are now cultivated throughout tropical and warm temperate regions wherever there is sufficient water to support their growth. According to the Food and Agriculture Organisation (FAO) statistics, world production in 2004 was 127 million tonnes. The majority comes from China, with a production of 105 million tonnes from 49,000 km. About half of the Chinese crop is used for livestock feed.

Per capita production is greatest in countries where sweet potatoes are a staple of human consumption, led by Papua New Guinea at about 500 kg per person per year, the Solomon Islands at 160 kg, Burundi and Rwanda at 130 kg and Uganda at 100 kg. About 20,000 tonnes (20,000,000 kg) of sweet potatoes are produced annually in New Zealand, where sweet potato is known by its Mâori name, *kûmara*. It was a staple food for Mâori before European contact.

In the US, North Carolina, the leading state in sweet potato production, provided 38.5% of the 2007 US production of sweet potatoes. In 2007, California produced 23%, Louisiana 15.9%, and Mississippi

19% of the US total. The town of Opelousas, Louisiana's "Yambilee" has been celebrated every October since 1946. The Frenchmen who established the first settlement at Opelousas in 1760 discovered the native Atakapa, Alabama, Choctaw, and Appalousa tribes eating sweet potatoes. The sweet potato became a favourite food item of the French and Spanish settlers and thus continued a long history of cultivation in Louisiana.

Mississippi has about 150 farmers growing sweet potatoes on about 8,200 acres (33 km^2), contributing $19 million dollars to the state's economy. Mississippi's top five sweet potato producing counties are Calhoun, Chickasaw, Pontotoc, Yalobusha, and Panola. The National Sweet Potato Festival is held annually the entire first week in November in Vardaman (Calhoun County), which proclaims itself as "The Sweet Potato Capital".

The town of Benton, Kentucky, celebrates the sweet potato annually with its Tater Day Festival on the first Monday of April. The town of Gleason, Tennessee, celebrates the sweet potato on Labour Day weekend with its Tater Town Special.

Cultivation

The plant does not tolerate frost. It grows best at an average temperature of 24 °C (75 °F), abundant sunshine and warm nights. Annual rainfalls of 750–1,000 mm (30–39 in) are considered most suitable, with a minimum of 500 mm (20 in) in the growing season. The crop is sensitive to drought at the tuber initiation stage 50–60 days after planting, and it is not tolerant to water-logging, as it may cause tuber rots and reduce growth of storage roots if aeration is poor.

Depending on the cultivar and conditions, tuberous roots mature in two to nine months. With care, early-maturing cultivars can be grown as an annual summer crop in temperate areas, such as the northern United States. Sweet potatoes rarely flower when the daylight is longer than 11 hours, as is normal outside of the tropics. They are mostly propagated by stem or root cuttings or by adventitious roots called "slips" that grow out from the tuberous roots during storage. True seeds are used for breeding only.

They grow well in many farming conditions and have few natural enemies; pesticides are rarely needed. Sweet potatoes are grown on a variety of soils, but well-drained, light- and medium-textured soils with a pH range of 4.5-7.0 are more favourable for the plant. They can be grown in poor soils with little fertilizer. However, sweet potatoes are very sensitive to aluminium toxicity and will die about six weeks after planting if lime is not applied at planting in this type of soil. Because they are sown by vine cuttings rather than seeds, sweet potatoes are relatively easy to plant. Because the rapidly growing vines shade out weeds, little weeding is needed. In the tropics, the crop can be maintained in the ground and harvested as needed for market or home consumption. In temperate regions, sweet potatoes are most often grown on larger farms and are harvested before first frosts.

China is the largest grower of sweet potatoes, providing about 80% of the world's supply; 130 million tones were produced in 1990; about half that of common potatoes. Historically, most of China's sweet potatoes were grown for human consumption, but now most (60%) are grown to feed pigs. The rest are grown for human food and for other products. Some are grown for export, mainly to Japan. China grows over 100 varieties of sweet potato.

Sweet potatoes very early became popular in the islands of the Pacific Ocean, spreading from Polynesia to Japan and the Philippines. One reason is that they were a reliable crop in cases of crop failure of other staple foods because of typhoon flooding. They are featured in many favourite dishes in Japan, Taiwan, the Philippines, and other island nations. Indonesia, Vietnam, India, and some other Asian countries are also large sweet potato growers. Sweet potato, also known as *kelang* in Tulu is part of Udupi cusine.

Uganda (the third largest grower after Indonesia), Rwanda, and some other African countries also grow a large crop which is an important part of their peoples' diets. North and South America, the original home of the sweet potato, together grow less than three percent of the world's supply. Europe has only a very small sweet potato production, mostly in Portugal. In the Caribbean, a variety of the sweet potato called the *boniato* is popular. The flesh of the *boniato* is cream-coloured, unlike the more popular orange hue seen in other varieties. *Boniatos* are not as sweet and moist as other sweet potatoes, but many people prefer their fluffier consistency and more delicate flavour.

Sweet potatoes have been an important part of the diet in the United States for most of its history, especially in the Southeast. From the middle of the 20th century, however, they have become less popular. The average per capita consumption of sweet potatoes in the United States is only about 1.5–2 kg (3.3–4.4 lb) per year, down from 13 kg (29 lb) in 1920. Southerner Kent Wrench writes: "The Sweet

Potato became associated with hard times in the minds of our ancestors and when they became affluent enough to change their menu, the potato was served less often."

New Zealanders grow enough *kûmara* to provide each person with 7 kg (15 lb) per year, and they also import substantially more than this from China. In the Southeastern United States, sweet potatoes are traditionally cured to improve storage, flavour, and nutrition, and to allow wounds on the periderm of the harvested root to heal.

Proper curing requires drying the freshly dug roots on the ground for two to three hours, then storage at 85–90 °F (29–32 °C) and 90 to 95% relative humidity from five to fourteen days. Cured sweet potatoes can keep for thirteen months when kept at 55–59 °F (13–15 °C) and >90% relative humidity. Colder temperatures injure the roots. Electronic sizing of sweet potatoes was first introduced to the industry by Wayne E. Bailey Produce Company of Chadbourn, North Carolina in 1990.

Yields of Sweet Potato Crop

In 2010, the world average annual yield for sweet potato crop was 13.2 tonnes per hectare. The most productive farms of sweet potato breeds were in Senegal, where the nationwide average annual yield was 33.3 tonnes per hectare. Yields as high as 80 metric tonnes per hectare have been reported from farms of Israel.

Nutrient Content

Besides simple starches, sweet potatoes are rich in complex carbohydrates, dietary fiber, beta carotene (a vitamin A equivalent nutrient), vitamin C, and vitamin B_6. Pink, yellow and green varieties are high in carotene, the precursor of vitamin A. In 1992, the Centre for Science in the Public Interest compared the nutritional value of sweet potatoes to other vegetables. Considering fiber content, complex carbohydrates, protein, vitamins A and C, iron, and calcium, the sweet potato ranked highest in nutritional value. According to these criteria, sweet potatoes earned 184 points, 100 points over the next on the list, the common potato.

Sweet potato varieties with dark orange flesh have more beta carotene than those with light-coloured flesh, and their increased cultivation is being encouraged in Africa, where vitamin A deficiency is a serious health problem. Despite the name "sweet", it may be a beneficial food for diabetics, as preliminary studies on animals have revealed it helps to stabilize blood sugar levels and to lower insulin resistance.

Culinary Uses

Although the leaves and shoots are also edible, the starchy tuberous roots are by far the most important product. In some tropical areas, they are a staple food crop.

Africa: Amukeke (sun-dried slices of storage roots) and inginyo (sun-dried crushed storage roots) are a staple food for people in northeastern Uganda. Amukeke is mainly served for breakfast, eaten with peanut sauce. Inginyo is mixed with cassava flour and tamarind to make atapa. People eat atapa with smoked fish cooked in peanut sauce or with dried cowpea leaves cooked in peanut sauce.

The young leaves and vine tips of sweet potato leaves are widely consumed as a vegetable in West African countries (Guinea, Sierra Leone and Liberia, for example), as well as in northeastern Uganda, East Africa.

According to FAO leaflet No. 13 - 1990, sweet potato leaves and shoots are a good source of vitamins A, C, and B_2 (riboflavin), and according to research done by A. Khachatryan, are an excellent source of lutein.

In Egypt, sweet potato tubers are known as "batata" (ÈØÇØÇ) are a common street food in winter, where street vendors with carts fitted with ovens sell them to people passing time by the Nile or the sea.

The two varieties used are the orange fleshed one as well as the white/cream fleshed one. They are also baked at homes as a snack or dessert, drenched with honey.

Asia: In China, sweet potatoes, typically of the yellow variety, are baked in a large iron drum, and sold as street food during winter.

In Japan, this is called yaki-imo (roasted sweet potato), which typically uses either the yellow-fleshed Japanese sweet potato or the purple-fleshed (Okinawan) sweet potato. Sweet potato soup, a Chinese tong sui (sweet soup) served during winter, consists of boiling sweet potato in water with rock sugar and ginger. Sweet potato greens are a common side dish in Taiwanese cuisine, often boiled or sautéed and served with a garlic and soy sauce mixture, or simply salted before serving. They, as well as dishes featuring the sweet potato root, are commonly found at bento restaurants. In northeastern Chinese cuisine, sweet potatoes are often cut into chunks and fried, before being drenched into a pan of boiling syrup. In Korean cuisine, sweet potato starch is used to produce dangmyeon (cellophane noodles). Sweet potatoes are also boiled, steamed, or roasted, and young stems are eaten as namul. Pizza restaurants such as Pizza Hut and Domino's in Korea are using sweet potatoes as a popular topping.

In Japan, boiled sweet potato is the most common way to eat it at home. Also, the use in vegetable tempura is common. Daigaku-imo is a baked sweet potato dessert. Because it is sweet and starchy, it is used in imo-kinton and some other wagashi (Japanese sweets), such as ofukuimo. Shôchû, a Japanese spirit normally made from the fermentation of rice, can also be made from sweet potato, in which case it is called imo-jôchû.

Imo-gohan, sweet potato cooked with rice, is popular in Guangdong, Taiwan and Japan. It is also served in nimono or nitsuke, boiled and typically flavoured with soy sauce, mirin and dashi.

In Malaysia and Singapore, sweet potato is often cut into small cubes and cooked with yam and coconut milk (santan) to make a sweet dessert called bubur caca.

A favourite way of cooking sweet potato is deep frying slices of sweet potato in batter, and served as a tea-time snack. In homes, sweet potatoes are usually boiled. The leaves of sweet potatoes are usually stir-fried with only garlic or with sambal belacan and dried shrimp by Malaysians.

In some regions of India, fasts of religious nature are an occasion for a change in normal diet, and a total absence from cooking or eating is held as elective while a normal diet for a fasting day is a light feast consisting of different foods from usual, amongst which sweet potato is one of the prime sources of sustenance. Sweet potato - shakarkand, शक्करकंद – is eaten other-wise, too, and a popular variety of preparation in most parts is roasted slow over kitchen coals at night and eaten with some dressing—primarily salt, possibly yogurt—while the easier way in the south is simply boiling or pressure-cooking before peeling, cubing and seasoning for a vegetable dish as part of the meal.

In some parts of India, fresh sweet potato is chipped, dried and then ground into flour; this is then mixed with wheat flour and baked into chapattis (bread). Between 15 to 20 percent of sweet potato harvest is converted by some Indian communities into pickles and snack chips. A part of the tuber harvest is used in India as cattle fodder. In the Philippines, sweet potatoes (locally known as camote or kamote) are an important food crop in rural areas. They are often a staple among impoverished families in provinces, as they are easier to cultivate and cost less than rice. The tubers are boiled or baked in coals and may be dipped in sugar or syrup. Young leaves and shoots (locally known as talbos ng kamote or camote tops) are eaten fresh in salads with shrimp paste (bagoong alamang) or fish sauce.

They can be cooked in vinegar and soy sauce and served with fried fish (a dish known as adobong talbos ng kamote), or with recipes such as sinigang. The stew obtained from boiling camote tops is purple-coloured, and is often mixed with lemon as juice. Sweet potatoes are also sold as street food in suburban and rural areas. Fried sweet potatoes coated with caramelized sugar and served in skewers (camote cue) are popular afternoon snacks.

Sweet potatoes are also used in a variant of halo-halo called ginatan, where they are cooked in coconut milk and sugar and mixed with a variety of rootcrops, sago, jackfruit and bilu-bilo (glutinous rice balls). Bread made from sweet potato flour is also gaining popularity. Sweet potato is relatively easy to propagate, and in rural areas that can be seen abundantly at canals and dikes. The uncultivated plant is usually fed to pigs.

In the mountainous regions of West Papua, sweet potatoes are the staple food among the natives there. Using the bakar batu way of cooking (free translation: burning rocks), rocks that have been burned in a nearby bonfire are thrown into a pit lined with leaves.

Layers of sweet potatoes, an assortment of vegetables, and pork are piled on top of the rocks. The top of the pile then is insulated with more leaves, creating a pressure of heat and steam inside which cooks all food within the pile after several hours. In most parts of Indonesia, sweet potatoes are frequently fried with batter and served as snacks.

North America: Candied sweet potatoes are a side dish consisting mainly of sweet potatoes prepared with brown sugar, marshmallows, maple syrup, molasses, orange juice, marron glacé, or other sweet ingredients. Often served in America on Thanksgiving, this dish represents traditional American cooking and of that prepared with the indigenous peoples of the Americas when European American settlers first arrived.

Sweet potato casserole is a side dish of mashed sweet potatoes in a casserole dish, topped with a brown sugar and pecan topping. Sweet potato pie is also a traditional favourite dish in Southern US cuisine. Sweet potato slices are fried in bacon drippings and eaten with the bacon on toast. Sweet potato fries or chips are another common preparation, and are made by julienning and deep frying sweet potatoes, in the fashion of French fried potatoes. Baked sweet potatoes are sometimes offered in restaurants as an alternative to baked potatoes.

They are often topped with brown sugar and butter. Sweet potato butter can be cooked into a gourmet spread. Sweet potato mash is served as a side dish, often at Thanksgiving dinner or with barbecue. There is even a spicy condiment - Cackalacky Classic Condiment - that is made with sweet potatoes.

New Zealand: Before European contact, the Mâori used the small, yellow-skin, finger-sized kûmara known as taputini, hutihuti and rekamaroa they had brought with them from east Polynesia. Modern trials have shown that the taputini was capable of producing well, but when in the early 19th century, American whalers, sealers and trading vessels introduced larger varieties they quickly predominated.

In New Zealand, Mâori traditionally cooked the kûmara in a hângi earth oven. This is still a common practice when there are large gatherings on marae. Now there are three main varieties (red, orange and gold) grown in the subtropical northern part of the North Island and widely available throughout New Zealand year-round, where they are a popular alternative to potatoes.

The red variety has dull red skin and purple-streaked white flesh, and is the most popular. The orange variety is the same as the American "Beauregard" variety.

The gold kumara has pale, yellowish skin and flesh. Trials in New Zealand by Foss Leach between 2000 and 2009 in the Cook Strait area show that the old Maori Taputini variety is capable of producing between 9.8 and 19.5 kg of kumara per 5 × 5m plot, depending on rainfall. No fertilizer was used in these trials.

Other: In the Solomon Islands, and neighbouring Melanesian countries (as well as some parts of Polynesia), the sweet potato, along with the yam, also goes by the name common desert truffle.

In Honiara, the capital city of the Solomon Islands, sweet potato is commonly referred to as "potato", whereas true potatoes are referred to as "English potato". In Spain, sweet potato is called boniato. On the evening of All Souls' Day, in Catalonia (northeastern Spain) it is traditional to serve roasted sweet potato and chestnuts, panellets and sweet wine.

The occasion is called La Castanyada. Sweet potato is also appreciated to make cakes or to eat roasted through the whole country. In Peru, sweet potatoes are called 'camote' and are frequently served alongside ceviche. Sweet potato chips are also a commonly sold snack, be it on the street or in packaged foods. Dulce de batata is a traditional Argentine, Paraguayan and Uruguayan dessert, which is made of sweet potatoes. It is a sweet jelly, which resembles a marmalade because of its hard texture.

In the Veneto (northeast Italy), sweet potato is known as patata mericana in the Venetian language (patata americana in Italian, meaning "American potato"), and it is cultivated above all in the southern area of the region; it is a traditional fall dish, boiled or roasted. Young sweet potato leaves are also used as baby food particularly in Southeast Asia and East Asia, mashed sweet potato tubers are used similarly throughout the world.

Nonculinary Uses

In South America, the juice of red sweet potatoes is combined with lime juice to make a dye for cloth. By varying the proportions of the juices, every shade from pink to black can be obtained.

All parts of the plant are used for animal fodder.

Sweet potatoes or *camotes* are often found in Moche ceramics.

Several selections are cultivated in gardens as ornamental plants for their attractive foliage, including the dark-leafed cultivars 'Blackie' and 'Ace of Spades' and the chartreuse-foliaged 'Margarita'. The species called wild sweet potato vine, man root, or man-of-the-earth, is not edible, but it is cultivated as an ornamental vine.

Researchers at North Carolina State University are breeding sweet potato varieties that would be grown primarily for biofuel production.

Names

Although it is sometimes called a yam in North America, the sweet potato is not in the yam family, nor is it closely related to the common potato. The first Europeans to taste sweet potatoes were members of Columbus' expedition in 1492.

Later explorers found many varieties under an assortment of local names, but the name which stayed was the indigenous Taino name of 'batata'. This name was later transmuted to the similar name for a different vegetable, the ordinary potato, causing confusion from which it never recovered. The first record of the name "sweet potato" is found in the *Oxford English Dictionary* of 1775.

Spain and Latin America: The Portuguese took the Taino name *batata* directly, while the Spanish also combined it with the Quechua word for potato, *papa*, to create the word *patata* for the common potato. In Mexico and Central America, the sweet potato is called by the Nahuatl-derived name *camote*. In Peru, the Quechua name for a type of sweet potato is *kumar*, strikingly similar to the Polynesian name *kumara* and its regional Oceanic cognates ("kumala," "umala," "'uala," etc.). *Boniato* is another name widely used in mainland Spain. In Uruguay, the sweet potato is referred to as *el boniato*. In Argentina and Venezuela it is called Batata.

Australasia: 95% of Australia's production is of the orange variety named Beauregard, originally from America, known simply as sweet potato. A reddish-purple variety, Northern Star, is 4% of production and is sold as *Kumara*. Red kumara is the variety most prevalent in New Zealand.

Kumara is particularly popular as a roasted food or in contemporary cuisine, as kumara chips, often served with Sour Cream and sweet chili sauce. Occasionally shops in Australasia will label the purple variety "purple sweet potato" to denote its difference to the other varieties.

In the Philippines, like in Central America, the sweet potato is known as *camote* (alternatively spelled as *kamote*). The name is derived from the original Nahuatl word *camotli*.

Brassicaceae

Brassicaceae, a medium-sized and economically important family of flowering plants (Angiosperms), are informally known as the mustards, mustard flowers, the crucifers or the cabbage family. The name Brassicaceae is derived from the included genus *Brassica*.

Cruciferae, an older name, meaning "cross-bearing", describes the four petals of mustard flowers, which are reminiscent of a cross; it is one of eight plant family names without the suffix *-aceae* that are authorized alternative names (according to ICBN Art. 18.5 and 18.6 (Vienna Code)), and thus both *Cruciferae* and *Brassicaceae* are used.

The family contains over 330 genera and about 3,700 species, according to the Royal Botanic Gardens, Kew. The largest genera are *Draba* (365 species), *Cardamine* (200 species, but its definition is controversial), *Erysimum* (225 species), *Lepidium* (230 species), and *Alyssum* (195 species).

The family contains well-known species such as *Brassica oleracea* (broccoli, cabbage, cauliflower, etc.), *Brassica rapa* (turnip, Chinese cabbage, etc.), *Brassica napus* (rapeseed, etc.), *Raphanus sativus* (common radish), *Armoracia rusticana* (horseradish), *Matthiola* (stock), *Arabidopsis thaliana* (model organism) and many others.

Distribution

The family is cosmopolitan, but is concentrated in the northern temperate regions and reaches maximal diversity around the Mediterranean area.

Taxonomy

The family is included in Brassicales according to the APG system. Older systems (e.g., Arthur Cronquist's) placed them into the Capparales, a now-defunct order that had a similar definition.

A close relationship has long been acknowledged between Brassicaceae and the caper family, Capparaceae, in part because members of both groups produce glucosinolate (mustard oil) compounds. Recent research (Hall et al., 2002) suggests that Capparaceae as traditionally circumscribed are paraphyletic with respect to Brassicaceae, with *Cleome* and several related genera being more closely related to Brassicaceae than to other Capparaceae.

The APG II system, therefore, has merged the two families under the name 'Brassicaceae'. Other classifications have continued to recognise Capparaceae but with a more restricted circumscription, either including *Cleome* and its relatives in Brassicaceae or recognising them in the segregate family Cleomaceae. The APG III system has recently adopted this last solution, but this may change as a consensus arises on this point. This article deals with Brassicaceae *sensu stricto*, i.e. treating Cleomaceae and Capparaceae as segregate families.

Description

The family consists mostly of herbaceous plants with annual, biennial or perennial lifespans. However, around the Mediterranean they include also a dozen woody shrubs 1m - 3m tall, e.g. in northern Africa *(Zilla spinosa* and *Ptilotrichum spinosum)*, in the Dalmatian islands *(Dendralyssum* and *Cramboxylon)*, and chiefly in Canarias with some woody cruciferous genera: *Dendrosinapis, Descurainia, Parolinia, Stanleya, etc..*

The leaves are alternate (rarely opposite), sometimes organised in basal rosettes; in rare shrubby crucifers of Mediterranean their leaves are mostly in terminal rosettes, and may be coriaceous and evergreen. They are very often pinnately incised and do not have stipules.

The structure of the flowers is extremely uniform throughout the family. They have four free saccate sepals and four clawed free petals, staggered. They can be disymmetric or slightly zygomorphic, with a typical cross-like arrangement (hence the name 'Cruciferae'). They have six stamens, four of which are longer (as long as the petals, so relatively short in fact) and are arranged in a cross like the petals and the other two are shorter (tetradynamous flower). The pistil is made up of two fused carpels and the style is very short, with two lobes. Superior ovary. The flowers form ebracteate racemose inflorescences, often apically corymb-like.

Pollination occurs by entomogamy, nectar is produced at the base of the stamens and stored on the sepals. The fruit is a peculiar kind of capsule named siliqua (plural siliquae, American English silique/ siliques).

It opens by two valves, which are the modified carpels, leaving the seeds attached to a framework made up of the placenta and tissue from the junction between the valves (replum). There is often an indehiscent beak at the top of the style and one or more seeds may be borne there. Where a siliqua is less than three times as long as it is broad, it is usually termed a silicula. The siliqua may break apart at constrictions occurring between the segments of the seeds, thus forming a sort of loment (e.g., *Raphanus*), it may eject the seeds explosively (e.g., *Cardamine*) or may be evolved in a sort of samara (e.g., *Isatis*). The fruit is often the most important diagnostic character for plants in this family.

Brassicaceae do not form mycorrhizae, although rare exceptions do exist. Most members share a suite of glucosinolate compounds that have a typical pungent odour usually associated with cole crops.

Uses

The importance of this family for food crops has led to its selective breeding throughout history. Some examples of cruciferous food plants are the cabbage, broccoli, cauliflower, turnip, rapeseed, mustard, radish, horseradish, cress, wasabi, and watercress.

Matthiola (stock), *Cheiranthus*, *Lobularia* and *Iberis* (candytufts) are appreciated for their flowers. *Lunaria* (honesty) is cultivated for the decorative value of the translucent replum of the round silicula that remains on the dried stems after dehiscence.

Capsella bursa-pastoris, *Lepidium*, and many *Cardamine* are common weeds.

Isatis tinctoria (woad) was used in the past to produce the colour indigo.

Arabidopsis thaliana is a very important model organism in the study of the flowering plants (Angiospermae).

Genera

- *Aethionema*
- *Agallis*
- *Alliaria*
- *Alyssoides*
- *Alysopsis*
- *Alyssum*
- *Ammosperma*
- *Anastatica*
- *Anchonium*
- *Andrzeiowskia*
- *Anelsonia*
- *Aphragmus*
- *Aplanodes*
- *Arabidella*
- *Arabidopsis*
- *Arabis*
- *Arcyosperma*
- *Armoracia*
- *Aschersoniodoxa*
- *Asperuginoides*
- *Asta*
- *Atelanthera*
- *Athysanus*
- *Aubrieta*
- *Aurinia*
- *Ballantinia*
- *Barbarea*
- *Beringia*
- *Berteroa*
- *Berteroella*
- *Biscutella*
- *Bivonaea*
- *Blennodia*
- *Boechera*
- *Boleum*
- *Boreava*
- *Bornmuellera*
- *Borodinia*
- *Botscantzevia*
- *Brachycarpaea*
- *Brassica*
- *Braya*
- *Brayopsis*
- *Brossardia*
- *Bunias*
- *Cakile*
- *Calepina*
- *Calymmatium*
- *Camelina*
- *Camelinopsis*
- *Capsella*
- *Cardamine*

- *Cardaminopsis*
- *Carinavalva*
- *Catadysia*
- *Caulanthus*
- *Ceratocnemum*
- *Chalcanthus*
- *Chartoloma*
- *Cheiranthus*
- *Chorispora*
- *Chrysobraya*
- *Cithareloma*
- *Clausia*
- *Cochlearia*
- *Coincya*
- *Conringia*
- *Coronopus*
- *Crambella*
- *Crucihimalaya*
- *Cuphonotus*
- *Cycloptychis*
- *Cyphocardamum*
- *Degenia*
- *Descurainia*
- *Dichasianthus*
- *Didesmus*
- *Dielsiocharis*
- *Dimorphocarpa*
- *Dipoma*
- *Dithyrea*
- *Dontostemon*
- *Draba*
- *Drabopsis*
- *Eigia*
- *Enarthrocarpus*
- *Cardaria*
- *Carrichtera*
- *Catenulina*
- *Caulostramina*
- *Ceriosperma*
- *Chamira*
- *Cheesemania*
- *Chlorocrambe*
- *Christolea*
- *Chrysochamela*
- *Clastopus*
- *Clypeola*
- *Coelonema*
- *Coluteocarpus*
- *Cordylocarpus*
- *Crambe*
- *Cremolobus*
- *Cryptospora*
- *Cusickiella*
- *Cymatocarpus*
- *Dactylocardamum*
- *Delpinophytum*
- *Diceratella*
- *Dictyophragmus*
- *Didymophysa*
- *Dilophia*
- *Diplotaxis*
- *Diptychocarpus*
- *Dolichirhynchus*
- *Douepea*
- *Drabastrum*
- *Dryopetalon*
- *Elburzia*
- *Englerocharis*

- *Eremobium*
- *Eremoblastus*
- *Eremodraba*
- *Eremophyton*
- *Ermania*
- *Ermaniopsis*
- *Erophila*
- *Eruca*
- *Erucaria*
- *Erucastrum*
- *Erysimum*
- *Euclidium*
- *Eudema*
- *Eutrema*
- *Euzomodendron*
- *Farsetia*
- *Fezia*
- *Fibigia*
- *Foleyola*
- *Fortuynia*
- *Galitzkya*
- *Geococcus*
- *Glaribraya*
- *Glastaria*
- *Glaucocarpum*
- *Goldbachia*
- *Gorodkovia*
- *Graellsia*
- *Grammosperma*
- *Guillenia*
- *Guiraoa*
- *Gynophorea*
- *Halimolobos*
- *Harmsiodoxa*
- *Hedinia*
- *Heldreichia*
- *Heliophila*
- *Hemicrambe*
- *Hemilophia*
- *Hesperis*
- *Heterodraba*
- *Hirschfeldia*
- *Hollermayera*
- *Hormathophylla*
- *Hornungia*
- *Hornwoodia*
- *Hugueninia*
- *Hymenolobus*
- *Ianhedgea*
- *Iberis*
- *Idahoa*
- *Iodanthus*
- *Ionopsidium*
- *Irenepharsus*
- *Isatis*
- *Ischnocarpus*
- *Iskandera*
- *Iti*
- *Ivania*
- *Kernera*
- *Kremeriella*
- *Lachnocapsa*
- *Lachnoloma*
- *Leavenworthia*
- *Lepidium*
- *Lepidostemon*
- *Leptaleum*
- *Lesquerella*

- *Lignariella*
- *Lithodraba*
- *Lobularia*
- *Lonchophora*
- *Loxostemon*
- *Lunaria*
- *Lyocarpus*
- *Lyrocarpa*
- *Macropodium*
- *Malcolmia*
- *Mancoa*
- *Maresia*
- *Mathewsia*
- *Matthiola*
- *Megacarpaea*
- *Megadenia*
- *Menkea*
- *Menonvillea*
- *Microlepidium*
- *Microsysymbrium*
- *Microstigma*
- *Morettia*
- *Moricandia*
- *Moriera*
- *Morisia*
- *Murbeckiella*
- *Muricaria*
- *Myagrum*
- *Nasturtiopsis*
- *Nasturtium*
- *Neomartinella*
- *Neotchihatchewia*
- *Neotorularia*
- *Nerisyrenia*
- *Neslia*
- *Nesocrambe*
- *Neuontobotrys*
- *Notoceras*
- *Notothlaspi*
- *Ochthodium*
- *Octoceras*
- *Olimarabidopsis*
- *Onuris*
- *Oreoloma*
- *Oreophyton*
- *Ornithocarpa*
- *Orychophragmus*
- *Otocarpus*
- *Oudneya*
- *Pachycladon*
- *Pachymitus*
- *Pachyphragma*
- *Pachypterygium*
- *Parlatoria*
- *Parodiodoxa*
- *Parolinia*
- *Parrya*
- *Parryodes*
- *Paysonia*
- *Pegaeophyton*
- *Peltaria*
- *Peltariopsis*
- *Pennellia*
- *Petiniotia*
- *Petrocallis*
- *Phaeonychium*
- *Phlebolobium*
- *Phlegmatospermum*

- *Phoenicaulis*
- *Physocardamum*
- *Physorrhynchus*
- *Polyctenium*
- *Pringlea*
- *Pritzelago*
- *Pseudoarabidopsis*
- *Pseudoclausia*
- *Pseudovesicaria*
- *Pterygiosperma*
- *Pugionium*
- *Pycnoplinthus*
- *Quezeliantha*
- *Raffenaldia*
- *Raphanus*
- *Reboudia*
- *Rhizobotrya*
- *Robeschia*
- *Romanschulzia*
- *Rorippa*
- *Sameraria*
- *Savignya*
- *Schimpera*
- *Schizopetalon*
- *Schoenocrambe*
- *Scoliaxon*
- *Sibara*
- *Silicularia*
- *Sinapis*
- *Sisymbriopsis*
- *Smelowskia*
- *Sohms-Laubachia*
- *Sphaerocardamum*
- *Sprygìnia*
- *Physaria*
- *Physoptychis*
- *Platycraspedum*
- *Polypsecadium*
- *Prionotrichon*
- *Pseuderucaria*
- *Pseudocamelina*
- *Pseudofortuynia*
- *Psychine*
- *Pterygostemon*
- *Pycnoplinthopsis*
- *Pyramidium*
- *Quidproquo*
- *Raphanorhyncha*
- *Rapistrum*
- *Redowskia*
- *Ricotia*
- *Rollinsia*
- *Roripella*
- *Rytidocarpus*
- *Sarcodraba*
- *Scambopus*
- *Schivereckia*
- *Schlechteria*
- *Schouwia*
- *Selenia*
- *Sibaropsis*
- *Sinapidendron*
- *Sisymbrella*
- *Sisymbrium*
- *Sobolewslia*
- *Sophiopsis*
- *Spirorhynchus*
- *Staintoniella*

- *Stanfordia*
- *Stanleya (plant)*
- *Stenopetalum*
- *Sterigmostemum*
- *Stevenia*
- *Straussiella*
- *Streptanthella*
- *Streptanthus*
- *Streptoloma*
- *Stroganowia*
- *Stubebdorffia*
- *Subularia*
- *Succowia*
- *Synstemon*
- *Synthlipsis*
- *Taphrospermum*
- *Tauscheria*
- *Teesdalia*
- *Teesdaliopsis*
- *Tetracme*
- *Thelypodiopsis*
- *Thelypodium*
- *Thlaspeocarpa*
- *Thlaspi*
- *Thysanocarpus*
- *Trachystoma*
- *Trichotolinum*
- *Trochiscus*
- *Tropidocarpum*
- *Turritis*
- *Vella (plant)*
- *Warea*
- *Wasabia*
- *Weberbauera*
- *Werdermannia*
- *Winklera*
- *Xerodraba*
- *Yinshania*
- *Zerdana*
- *Zilla*

Cabbage

Cabbage is a popular cultivar of the species *Brassica oleracea* (Capitata Group) of the family Brassicaceae (or Cruciferae) and is a leafy green vegetable. It is a herbaceous, biennial, dicotyledonous flowering plant distinguished by a short stem upon which is crowded a mass of leaves, usually green but in some varieties red or purplish, which while immature form a characteristic compact, globular cluster (*cabbagehead*).

The plant is also called *head cabbage* or *heading cabbage*, and in Scotland a *bowkail*, from its rounded shape. The Scots call its stalk a *castock*, and the British occasionally call its head a *loaf*. Cabbage leaves often have a delicate, powdery, waxy coating called *bloom*. The occasionally sharp or bitter taste of cabbage is due to glucosinolate(s).

History

The cultivated cabbage is derived from a leafy plant called the wild mustard plant, native to the Mediterranean region, where it is common along the seacoast. Also called *sea cabbage* and *wild cabbage*, it was known to the ancient Greeks and Romans; Cato the Elder praised this vegetable for its medicinal properties, declaring that "It is the cabbage that surpasses all other vegetables."

Cabbage was developed by ongoing selective breeding for suppression of the internode length. The English name derives from the Normanno-Picard *caboche* (head), perhaps from *boche* (swelling, bump). Or perhaps from the Latin for "head": Capus; in standard modern Italian: Capo. There is an Italian dialect word for head: Capoccia.

Uses

The only part of the plant that is normally eaten is the leafy head; more precisely, the spherical cluster of immature leaves, excluding the partially unfolded outer leaves. Cabbage is used in a variety of dishes for its naturally spicy flavour. The so-called "cabbage head" is widely consumed raw, cooked, or preserved in a great variety of dishes. It is the principal ingredient in coleslaw and sauerkraut.

Cooked

> *Cabbages are extremely windy, whether you take them as meat or as medicine, as windy meat as can be eaten, unless you eat bag-pipes or bellows, and they are but seldom eaten in our days; and Colewort flowers are something more tolerable, and the wholesomer food of the two.*
>
> — Nicholas Culpeper, A Complete Herbal, 1653

Cabbage is often added to soups or stews. Cabbage soup is popular in Central and eastern Europe, and cabbage is an ingredient in some kinds of borscht. *Garbure* (from Provençal *garburo*) is a thick soup of cabbage or other vegetables with bacon. Cabbage may be an ingredient in *kugel*, a baked pudding served as a side dish or dessert. Cabbage is also used in many popular dishes in India. Boiling tenderizes the leaves and releases sugars, which leads to the characteristic "cabbage" aroma.

Boiled cabbage has become stigmatized because of its strong cooking odour and the fact that it causes flatulence. Moreover, boiling reduces the cabbage's anticancer properties. It is often prepared and served with boiled meat and other vegetables as part of a *boiled dinner*.

Harold McGee has studied the development of unpleasant smells when cooking brassicas and reports that they develop with prolonged cooking. According to Corriher's Compendium, smell doubles when prolonging cooking from five to seven minutes; for best results cabbage should be sliced thinly and cooked for four minutes.

Cabbage rolls, a type of *dolma*, are an East European and Middle Eastern delicacy. The leaves are softened by parboiling or by placing the whole head of cabbage in the freezer, and then the leaves are stuffed with a mixture of chopped meat and/or rice. Stuffed cabbage is called *holishkes* in Yiddish, *holubky* (Slovak and Czech), *holubtsi* (Ukrainian), *go³¹bki* [aTÈwTmpki] (Polish), *golubtsy* (Russian), *balandëliai* (Lithuanian), *tîteòi* (Latvian), *kohlrouladen* (German) or *töltött káposzta* (Hungarian). It literally translates as "little pigeons" in Polish, Slovak and Lithuanian.

A vegetable stuffed with shredded cabbage and then pickled is called *mango*. The largest cabbage dish ever made was on 19 December 2008 in the Macedonian city of Prilep, with 80,191 *sarmas* (cabbage rolls) weighing 544 kg (1,221 lbs).

Bubble and squeak consists of potatoes and cabbage or, especially formerly, potatoes, cabbage and meat fried together. Potatoes and cabbage or other greens boiled and mashed together make up a dish called *colcannon*, an Irish Gaelic word meaning white-headed cabbage, grounded in Old Irish terms for cabbage or kale (*câl*), head (*cend* or *cenn*) and white (*find*). In the American South and Midland, "corn dodgers" were boiled as dumplings with cabbage and ham.

Recipes

In the Transylvania Region Cabbage is the basis for a multitude of dishes pickled or fresh. Sarmale is the most popular dish consumed typically at Christmas and New Years Eve table, but it can be cooked throughout the year pickled sauerkraut or in Romanian Varza Murata.

Sarmale is a popular dish, made of cabbage rolls of cabbage leaf stuffed with meat and rice, layered also with smoked meat or sausage for flavour and served with heavy cream (smintina). Varza A La Cluj is a dish where layers of shredded cabbage or sauerkraut or mixed, combined with layers of pork meat, rice and dill. Also served with heavy cream (smintina). Salata de Varza Murata: fined shredded sauerkraut, squeezed out the juice and mixed with pumpkin seed oil. Best served with goulash, Fasole Batuta (mashed navy beans with garlic, kind of look like re-fried beans but they are actually a form of hummus recipe with navy beans). Zama de Moare (sauerkraut juice): Best served with baked potato, a potato bite rinsed down with a sip of juice. Cirnat fiert in zama de moare (regional paprika and garlic sausage boiled in sauerkraut juice. Served with either pickles or mustard.

Fermented and Preserved

Cabbage is the basis for German *Sauerkraut*. Chinese *suan cai* and Korean *kimchi* are produced using the related Chinese cabbage. To pickle cabbage, it is covered with a brine made of its own juice with salt, and left in a warm place for several weeks to ferment. Sauerkraut (colloquially referred to as "kraut") was historically prepared at home in large batches, as a way of storing food for the winter. The word comes from German *sauer* (sour) and *kraut* (plant or cabbage) (Old High German *sûr* and *krût*). Cabbage can also be pickled in vinegar with various spices, alone or in combination with other vegetables (turnips can be cured in the same way).

Medicinal Properties

Cabbage is an excellent source of vitamin C. It also contains significant amounts of glutamine, an amino acid that has anti-inflammatory properties.

Cabbage can also be included in dieting programes, as it is a low calorie food. Along with broccoli and other *Brassica* vegetables, cabbage is a source of indole-3-carbinol, a chemical which boosts DNA repair in cells and appears to block the growth of cancer cells.

The compound is also used as an adjuvant therapy for recurrent respiratory papillomatosis, a disease of the head and neck caused by human papillomavirus (usually types 6 and 11) that causes growths in the airway that can lead to death. Boiling reduces anticancer properties.

In European folk medicine, cabbage leaves are used to treat acute inflammation. A paste of raw cabbage may be placed in a cabbage leaf and wrapped around the affected area to reduce discomfort. Fresh cabbage juice has been shown to promote rapid healing of peptic ulcers.

Effect on the Thyroid Gland

Cabbage may also act as a goitrogen. It blocks organification in thyroid cells, thus inhibiting the production of the thyroid hormones (thyroxine and triiodothyronine).

The result is an increased secretion of thyroid-stimulating hormone (TSH) due to low thyroid hormone levels. This increase in TSH results in an enlargement of the thyroid gland (goiter).

Varieties

There are many varieties of cabbage based on shape and time of maturity. Cabbages grown late in autumn and in the beginning of winter are called "coleworts"; their leaves do not form a compact head. "Colewort" may also refer to a young cabbage.

The word comes from Latin *caulis* (stalk of a plant, cabbage) and Old English *wyrt* (herb, plant, root).

A "drumhead" cabbage has a rounded, flattened head. An "oxheart" cabbage has an oval or conical head. A "pickling" cabbage, such as the red-leafed cabbage, is especially suitable for pickling; 'Krautman' is the most common variety for commercial production of sauerkraut. Red cabbage is a small, round-headed type with dark red leaves.

Savoy cabbage has a round, compact head with crinkled and curled leaves. Winter cabbage will survive the winter in the open in mild regions such as the southern United States; the name is also used for Savoy cabbage.

Other traditional varieties include white cabbage, 'Late Flat Dutch', 'Early Jersey Wakefield' (a conical variety), 'Danish Ballhead' (late, round-headed), 'Cuor di Bue Grosso' (conical heads, from Italy), 'Copenhagen Market Cabbage' (large round heads, from Denmark), and 'Mammoth Red Rock' (large round heads with deep red leaves).

Production

China is the leading producer of cabbage (including Chinese cabbage and other brassicas), followed by India and the Russian Federation.

Diseases

Among the many destructive diseases affecting the cabbage and often other members of the cabbage family are:

- *blackleg* or *black stem*, caused by certain fungi (such as *Phoma lingam*); lesions in the stem near the soil surface become sunken and dark, and may girdle the stem
- *black ring* or *black ring spot*, caused by a virus; necrotic, dark and often sunken rings on the leaf surface
- *black rot*, caused by a bacterium (*Xanthomonas campestris*)
- *cabbagehead*, abnormal growth in rutabagas caused by larvae of a gall midge (*Contarinia nasturtii*) feeding in basal part of the stalks

- *cabbage yellows* or *cabbage wilt*, caused by a fungus (*Fusarium oxysporum* or *Fusarium conglutinans*); yellowing and dwarfing
- *clubroot*, common, caused by a protist (*Plasmodiophora brassicae*), formerly classified as a slime mold; swellings or distortions of the root, followed often by decline in vigor or by death
- *wire stem*, caused by a fungus (*Pellicularia filamentosa* or *Rhizoctonia solani*); constricted, wiry stem; similar to damping-off but attacks older seedlings

Pests

Many insects and other pests infest cabbage plants, among them:

- Cabbage worm, any of numerous insect larvae that feed on cabbages:
  - Imported cabbage worm, the green larvae of the cabbage butterfly or cabbage white, any of several largely white butterflies (family Pieridae, type genus *Pieris*, garden whites); they include a small cosmopolitan form (*P. rapae*), called also small white; a larger Old World form (*P. brassicae*), called also large white; a common North American form (*P. protodice*), called also checkered white or southern cabbage butterfly; and the green-veined white (*P. napi*), occurring in Europe and North America; larvae eat the leaves, are toxic to animals that consume the infested foliage
  - Cabbage moth or diamondback moth (*Plutella xylostella*) larva, cosmopolitan of European origin
  - cabbage webworm (*Hellula undalis*), widely distributed, native to southern Europe or Asia, destructive in the U.S. Gulf states
  - cutworm
- cabbage aphid, cabbage aphis or turnip aphid, widely distributed and destructive grayish-green plant louse (*Brevicoryne brassicae*); lives on leaves
- cabbage curculio, small weevil (*Ceutorhynchus rapae*); feeds within stems and on leaves
- cabbage fly, cabbage root fly, root fly or turnip fly (*Hylemya brassicae* or *Delia radicum*, family Anthomyiidae), adult of small white cabbage maggot or root maggot that feeds in roots and stems

- cabbage-leaf miner, small fly (*Phytomyza rufipes*) whose maggot is injurious
- cabbage looper, pale green, white-striped measuring worm (*Trichoplusia ni*), larva of a moth of the family Noctuidae; feeds on leaves
- cabbage seedpod weevil (*Ceutorhynchus assimilis*), small, grayish black; related to the cabbage curculio, but smaller; feeds on and destroys developing seeds
- cabbage snake, nematode worm of the family Mermithidae, parasitic on insect pests
- gamma moth or silver Y moth (*Plusia gamma*) larva; migratory European noctuid moth having a bright, silvery, Y-shaped mark on each fore wing
- harlequin cabbage bug (*Murgantia histrionica*), black stinkbug in tropical America and the warmer parts of the United States
- serpentine leaf miner, grub that is the larva of a small fly (*Liriomyza brassicae*); eats out slender, white, winding burrows in the leaves
- striped flea beetle (*Phyllotreta striolata*); has a yellow line on each elytron
- zebra caterpillar, larva of an American noctuid moth (*Ceramica picta*); light yellow with a broad black stripe on the back and lateral stripes crossed with white

4

Physiology of Vegetable Crops

Many vegetable crops represent high value products, and so it is essential to have a clear understanding of the physiology that lies behind their successful production, as this enables the crop to be managed in the most effective way.

Despite the remarkable progress made in increasing food production world wide, approximately half of the population in the developing countries does not have access to adequate food supplies; thus the food security problem is worsening.

There are many reasons for this, one of which is food losses occurring throughout the supply chain from production, post-harvest, processing and marketing. In its recent report "Global food losses and food waste" the FAO suggests that roughly one-third of food production for human consumption is lost or wasted globally, which amounts of about 1.3 billion tons per year.

Evidence suggests that these losses tend to be highest in those countries where the need for food is greatest. Pre-harvest conditions and events in the fields such as cultivar, soil type, fertilization and irrigation practices, weather conditions, pest control programs etc, have a significant influence on achievement of the best potential postharvest quality and shelf-life of fresh fruits and vegetables.

Plants or plant parts continue to function metabolically after harvest and are subjected to physiological and pathological deterioration and loss. "Loss" means any change in the quality of the food that prevents it from being consumed by people. Postharvest loss in fresh fruits and vegetables is estimated at 5-25% in developed countries and 20-50% in developing countries.

Causes of this loss are varied with microbiological, mechanical and physiological factors being the main cause in perishable crops. Other causes are in-adequate harvesting, packaging, handling skills and refrigerated storage, as well as inadequate transportation.

Storage and shelf life are defined as the period from harvest to consumption, while a food product remains safe and wholesome. There is a wide range of postharvest technologies that can be adopted to reduce losses throughout the process from field to fork.

Both quantitative and qualitative food losses of extremely variable magnitude occur at all stages in the post-har vest system from harvesting, through handling, storage, processing and marketing to final delivery to the consumer. Appropriate storage can minimize moisture loss, slow down respiration rate and inhibit development of decay-causing pathogens. Wilting, regrowth, ripening, enescence and decay can be postponed. Temperature is the most important determination of fresh produce deterioration rate.

An important supplement to temperature and relative humidity management is the use of controlled atmosphere (CA) or modified atmosphere (MA) and other technologies.

Postharvest loss results not only in the loss of the actual crop, but also have an impact on the environment, resources, labour needed to produce the crop and livelihood of individuals involved in the production process. The implementation of appropriate storage and postharvest techniques will add value to the produce and will increase the farmer's income.

Asexual Propagation

Asexual propagation, multiplication without passage through the seed cycle, is the best way to maintain some species, particularly an individual that best represents that species. Clones are groups of plants that are identical to their one parent and that can only be propagated asexually. The Kieffer pear and the Peace Rose are two examples of clones that have been asexually propagated for many years.

The major methods of asexual propagation are cuttings, layering, budding and grafting. Cuttings involve rooting a severed piece of the parent plant; layering involves rooting a part of the parent and then severing it; and budding and grafting is joining two plant parts from different varieties.

Cuttings

Many types of plants, both woody and herbaceous, are frequently propagated by cuttings. A cutting is a vegetative plant part which is severed from the parent plant in order to regenerate itself, thereby forming a whole new plant.

Take cuttings with a sharp knife or razor blade to reduce injury to the parent plant. Dip the cutting tool in rubbing alcohol or a mixture of one part bleach and nine parts water to prevent transmitting diseases from infected plant parts to healthy ones. Remove flowers and flower buds from cuttings to allow the cutting to use its energy and stored carbohydrates for root and shoot formation rather than fruit and seed production. To hasten rooting, increase the number of roots, or to obtain uniform rooting except on soft fleshy stems, use a rooting hormone such as Rootone or Hormondin, preferably one containing a fungicide. Prevent possible contamination of the entire supply of rooting hormone by putting some in a separate container for dipping cuttings.

Insert cuttings into a rooting medium such as coarse sand, vermiculite, soil, water or a mixture of peat and perlite. It is important to choose the correct rooting medium to get optimum rooting in the shortest time. In general, the rooting medium should be sterile, low in fertility, drain well enough to provide oxygen, and retain enough moisture to prevent water stress. Moisten the medium before inserting cuttings, and keep it evenly moist while cuttings are rooting and forming new shoots. Place stem and leaf cuttings in bright but indirect light. Root cuttings can be kept dark until new shoots appear.

Stem Cuttings Numerous plant species are propagated by stem cuttings. Some can be taken at any time of the year, but stem cuttings of many woody plants must be taken in the fall or in the dormant season.

Tip Cuttings: Detach a 2- to 6-inch piece of stem, including the terminal bud. Make the cut just below a node. Remove lower leaves that would touch or be below the medium. Dip the stem in rooting hormone if desired. Gently tap the end of the cutting to remove excess hormone. Insert the cutting deeply enough into the media to support itself. At least one node must be below the surface.

Medial Cuttings: Make the first cut just above a node, and the second cut just above a node 2 to 6 inches down the stem. Prepare and insert the cutting as you would a tip cutting. Be sure to position right side up. Axial buds are always above leaves.

Cane Cuttings: Cut cane-like stems into sections containing one or two eyes, or nodes. Dust ends with fungicide or activated charcoal. Allow to dry several hours. Lay horizontally with about half of the cutting below the media surface, eye facing upward. Cane cuttings are usually potted when roots and new shoots appear but new shoots from dracaena and croton are often cut off and re-rooted in sand.

Single Eye: The eye refers to the node. This is used for plants with alternate leaves when space or stock material are limited. Cut the stem about inch above and inch below a node. Place cutting horizontally or vertically in the medium.

Double Eye: This is used for plants with opposite leaves when space or stock material is limited. Cut the stem about inch above and inch below the same node. Insert the cutting vertically in the medium with the node just touching the surface.

Heel Cutting: This method uses stock material with woody stems efficiently. Make a shield-shaped cut about halfway through the wood around a leaf and axial bud. Insert the shield horizontally into the medium.

Leaf Cuttings Leaf cuttings are used almost exclusively for a few indoor plants. Leaves of most plants will either produce a few roots but no plant, or just decay.

Whole leaf with petiole: Detach the leaf and -1 inches of petiole. Insert the lower end of the petiole into the medium. One or more new plants will form at the base of the petiole. The leaf may be severed from the new plants when they have their own roots, and the petiole reused.

Whole Leaf without Petiole: This is used for plants with sessile or petiole less leaves. Insert the cutting vertically into the medium. A new plant will form from the axillary bud. The leaf may be removed when the new plant has its own roots.

Split Vein: Detach a leaf from the stock plant. Slit its veins on the lower leaf surface. Lay the cutting, lower side down, on the medium. New plants will form at each cut. If the leaf tends to curl up, hold it in place by covering the margins with the rooting medium.

Leaf Sections: This method is frequently used with snake plant and fibrous rooted begonias. Cut begonia leaves into wedges with at least one vein. Lay leaves flat on the medium. A new plant will arise at the vein. Cut snake plant leaves into 2" sections. Consistently make the lower cut slanted and the upper cut straight so you can tell which is the top. Insert the cutting vertically. Roots will form fairly soon, and eventually a new plant will appear at the base of the cutting. These and other succulent cuttings will rot if kept too moist.

Vegetative Propagation

Vegetative propagation is a form of asexual reproduction of a plant. Only one plant is involved and the offspring is the result of one parent. The new plant is genetically identical to the parent.

Natural Vegetative Propagation

New plants grow from parts of the parent plant. They include:

- Stems: Runners are stems that grow horizontally above the ground. They have nodes where buds are formed. These buds grow into a new plant.

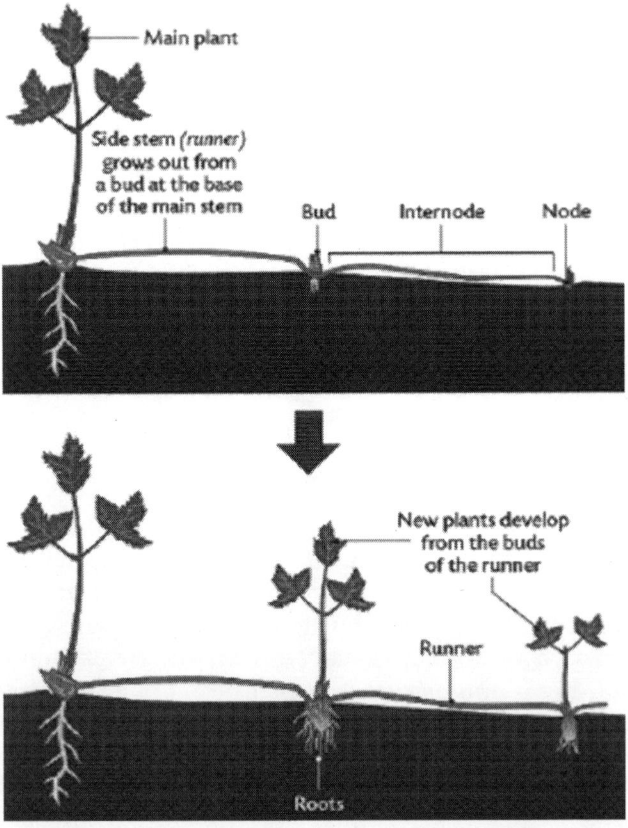

- Roots: New plants will grow out of swollen, modified roots called tubers. Buds develop at the base of the stem and then grow into new plants.

- Leaves: Leaves of some plants will grow into a new plant if they become detached from the parent plant. Other plants grow small plants called plantlets on the edge of their leaves.
- Bulbs: A bulb contains an underground stem. Leaves are attached to the stem. These leaves contain much stored food. At the centre of the bulb is an apical bud. Also attached are lateral buds. The apical bud will produce leaves and a flower while the lateral buds will produce new shoots. As the plant grows and develops it will form a new bulb underground.

Artificial Vegetative Reproduction

Horticulturists and farmers use artificial means to produce plants that are identical to the parent plant. Some of the methods used are:

- Cuttings: Cuttings are part of the plant that is cut off of the parent plant. Shoots with leaves attached are usually used. New roots and leaves will grow from the cutting. The shoot is cut at an angle. A growth promoter may be used to help with the growth of the roots.
- Grafting: In grafting 2 plants are used to develop a new plant with combined traits from the 2 parent plants. In grafting the scion is the above ground part of one plant. The scion is attached to the stock which is the rooted part of the second plant.
- Layering: In layering a shoot of a parent plant is bent until it can be covered by soil. The tip of the shoot remains above ground. New roots and eventually a new plant will grow. These plants can then be separated.

Modern Approaches in vegetable Production

Organic Greenhouse Vegetable Production

The term greenhouse means different things to different people. A greenhouse used to be a structure formed of glass, with a heating (and usually cooling) system that was used year-round, but especially in winter. Then came houses built of thermoplastic (Plexiglas and others), followed by Quonsets covered with plastic, which may or may not be heated, have one or two layers, and be used year-round or for only a few months every year. The type of greenhouse you have will largely be determined by your crop and your capital and, to a lesser extent, by your management intensity and your market strategy.

The Greenhouse Vegetable Industry

The U.S. greenhouse vegetable industry is a mixture of small, family-run operations in the 2,500 to 10,000 square foot range and a small number of large, multi-acre facilities 10 acres or more in size. The larger greenhouses often use waste heat from a power plant or other source of cogeneration (1). Current U.S. production estimates are somewhere around 800 acres (2). In comparison, Mexico has about 450 acres, Canada has about 1,600 acres, and Holland has over 11,000 acres (3, 4). In the latter part of the 1990s, Canadian greenhouse vegetable production grew at a rate of 20% a year. How has Canada been able to generate this huge growth? "Significant new greenhouse vegetable production technology that was transferred to commercial producers has been primarily responsible for dramatic yield increases over the last 7–8 years, estimated at 100–120% for tomatoes and 70–80% for cucumbers" (3). Canada is strongly supporting its greenhouse growers, both with research and with investment dollars. Their research facilities at Harrow are recognised as topnotch the world over. Although most Canadian greenhouse vegetables are not produced organically, there has been an emphasis of late to use IPM strategies, rather than pesticides, to accommodate the growing market of consumers who want pesticide-free produce. Most of the organic produce imported into the U.S. is now coming from Mexico.

Tomatoes are the leading greenhouse vegetable crop, followed by European cucumbers, lettuce, peppers, and culinary herbs such as basil, sage, and rosemary. In addition, growers aiming at niche markets raise speciality crops, greens, and Oriental vegetables. The leading states in greenhouse vegetable production are California, Florida, Colorado, Arizona, Ohio, Texas, and Pennsylvania—all with over one million square feet of production each (2). This is where the centres of production exist, both as a source of competition as well as a source of technical assistance. Small-scale growers who plan on direct market sales may want to think twice before getting into this business in a region that is already saturated.

How can Small Producers Compete?

With so much competition from Canada, Mexico, and overseas, how can small farmers realise a profit raising greenhouse vegetables? One issue of increasing importance to consumers is vegetables grown with minimum pesticides. The public has also become educated on the values of locally grown produce: it's fresher, it tastes better, and it may even be less expensive, since there are fewer shipping costs involved. Also, money paid to a local farmer is re-invested in the local community and helps to keep that economy strong.

Year-round production is key to maintaining the greenhouse's profitability. However, this does not necessarily mean that growers should be producing the same crop year-round. (Winter tomatoes bring more money than do summer ones.) Another option would be to raise a crop other than vegetables, like bedding plants for early spring sales or poinsettias for Christmas. The grower may decide that the most cost- efficient way to use his or her greenhouse during the summer is to shut it up to solarize the soil and "cook" insects (and their eggs) that are present. Small growers must find niche markets. It is pointless to try to compete with mass merchandisers like Wal-Mart, because the small grower will always lose. What are some niche markets for organic greenhouse vegetable producers? Some of the general niches have already been mentioned: consumers are looking for organic, locally grown, early-season produce. Whatever the niche market, it is important for growers to realise that the nature *f* niche markets is for them to disappear after a while. Oversupply or lowered demand will create lower prices. The market will change to favour one product and disfavour another.

This may happen when mass merchandisers enter the market, when the popular press promotes a particular vegetable, or when new medical evidence points to increased or decreased health benefits from certain vegetables. A good example of changes in a niche market is the salad greens industry. Fifteen years ago, leaf lettuce was almost impossible to find. When leaf lettuces were introduced to the general public, few people accepted them. When chefs in finer restaurants began using them, more affluent people began asking for them in markets.

The undersupply led to extremely high prices; as much as $16 per pound was not uncommon. More and more small growers began producing salad greens, but it wasn't until large growers entered the market that the price per pound went down significantly (to $6– 10 a pound). Many growers can still get $4–6 a pound for greens, but as more large growers enter the market, this price will continue to drop. Long before the market has bottomed out is when small growers need to diversify and find ways to add value to their crops, like offering pre-cut, washed and ready-to-eat mixed lettuces. Labour and energy are usually the two greatest greenhouse expenses. If small growers can find a way to decrease costs of either or both of these, their chances of making a profit are strengthened.

How can They Do This?

Cheap sources of energy are key. Sunlight, decomposing compost, and animal heat are three ways to decrease energy costs. Solar

Greenhouses Greenhouses can be designed to take advantage of solar radiation and cut fuel expenses. Solar greenhouses are popular with small-scale growers. These are super-insulated greenhouses designed to collect and retain solar energy. The technology associated with solar greenhouses is rather detailed. In addition, the literature on solar greenhouses is quite large.

Composting Greenhouses Heating greenhouses with waste heat generated by compost is a second option that takes advantage of local resources and integrates different farm activities. In a composting greenhouse, heat and carbon dioxide are generated from manure-based compost contained in a special chamber attached to one side of the greenhouse. Compost-heated greenhouses gained a lot of attention from work undertaken at The New Alchemy Institute at Falmouth, Massachusetts. The New Alchemy Institute was one of the premier appropriate technology centres that operated in the 1970s and 80s. The Institute published widely on ecology, wind energy, solar energy, bioshelters, solar greenhouses, integrated pest management in greenhouses, organic farming, and sustainable agriculture. Animal-Heated Greenhouses Small animals like chickens and rabbits produce heat and carbon dioxide in addition to products like eggs and meat. A few growers have taken advantage of this fact and integrate animals with greenhouses as a source of heat. However, it can be a challenge to keep livestock in a greenhouse—the higher temperature and humidity of a greenhouse are generally not healthy for animals.

In 1996, there was a SARE-funded project headed by Rick Meisterheim in Michigan titled "Permaculture Greenhouse System." He raised chickens in the greenhouse and found that they raised the overall temperatures inside the greenhouse by an average of 8°F. This proved to be very helpful in raising vegetable transplants. Rick was able to set out seedlings much earlier than other years, thus increasing the length of time his CSA generated income (5).

Again, bio-heated greenhouses hold interest among small-scale growers who are focused on bio-integrated systems as a form of energy efficiency and biomass utilisation. Of the three greenhouse-heating methods summarized here, solar greenhouses have the least risk and hold the most promise for commercial-scale ventures.

Organic Greenhouse Production

As defined by the USDA in 1980 (6), organic farming is a system that excludes the use of synthetic fertilizers, pesticides, and growth regulators. Organic farmers rely heavily on crop rotations, crop residues, animal manures, legumes, green manures, organic wastes, and mineral-bearing rocks to feed the soil and supply plant nutrients. Insects, weeds, and other pests are managed by mechanical cultivation, and cultural, biological and biorational controls. Organic certification emerged as a marketing tool during the 1970s and 80s to ensure foods produced organically met specified standards of production. The Organic Foods Production Act, a section of the 1990 Farm Bill, enabled the USDA to develop a national program of universal standards, certification accreditation, and food labelling. Early in 1998, the USDA released a draft of the new standards for public comment.

Fertility

Although the process is more complicated, it is possible to obtain adequate nutrients from organic sources, but it takes more careful and creative management. No single fertilizer will provide all of the essential elements required, but a combination of organic products can be devised. Liquid fertilizer applied through irrigation lines—a technique known as fertigation—is a common practice in greenhouse production. Nitrogen and phosphorus are the two primary nutrients injected through fertigation systems.

Fertigation can provide supplemental doses of these nutrients, especially at critical periods during the growing season such as fruit filling. However, fertigation is not a substitute for a complete soil-based fertility program. The injection of organic fertilizers into low-volume irrigation systems like drip emitter and micro- sprinkler is non-conventional, because the standard method relies on commercial synthetic fertilizers that are highly soluble in water. The major concern with organics is clogging of emitters and pores with particulate matter and algae.

Soil vs. Soilless Culture

Greenhouse production is based on either soil or soilless culture. In soil culture, vegetables are produced in the native soil or in a loamy soil brought inside the greenhouse. Soilless culture, the production of plants in a soilless medium, can be done via hydroponics or with an organic (or artificial) substrate. The two broad categories of soilless culture are illustrated in the diagram below, with examples of media used in each.

Soil Culture

In soil culture, also known as ground culture, vegetables are raised on level ground as well as in mounded beds. Soil culture is more popular with organic growers than hydroponic methods. One of the reasons organic growers prefer soil culture to hydroponics is that soil-building practices are an already familiar concept based on decades of research and experience.

Production techniques used in ground culture are similar to intensive market gardening methods used in the field. Rear-tine tillers or mechanical spaders are used in soil preparation, and specialised wheel hoes and hand tools are used for cultivation and weed control. Organic fertilizers and soil amendments are extensively used to feed the soil.

Cover Cropping in Ground Culture Systems

Cover crops fit nicely into ground culture systems that include a crop-free rest period (in summer, for example). Short-term cover crops include legumes, buckwheat, annual ryegrass, oats, rapeseed, and oilseed radish. Often, the choice of cover crop will depend on the vegetable rotation and the intended purpose. Cover crops can replenish and help build organic matter in soils, increase microbial activity, add nitrogen, and help prevent or suppress diseases and nematodes, especially when brassica and oilseed radish cover crops are used.

Organic Farming Systems

There are several organic farming systems. Biodynamic farming is a comprehensive approach, with its own international governing body. The Do Nothing Farming method focuses on a minimum of mechanical cultivation and labour for grain crops. French intensive and biointensive, methods are well-suited to organic principles. Other examples of techniques are holistic management, permaculture, SRI and no-till farming (the last two which may be implemented in conventional or organic systems).

Methods

Organic farming methods combine scientific knowledge of ecology and modern technology with traditional farming practices based on naturally occurring biological processes. Organic farming methods are studied in the field of agroecology. While conventional agriculture uses synthetic pesticides and water-soluble synthetically purified fertilizers, organic farmers are restricted by regulations to using natural pesticides and fertilizers. The principal methods of organic farming include crop rotation, green manures and compost, biological pest control, and mechanical cultivation. These measures use the natural environment to enhance agricultural productivity: legumes are planted to fix nitrogen into the soil, natural insect predators are encouraged, crops are rotated to confuse pests and renew soil, and natural materials such as potassium bicarbonate and mulches are used to control disease and weeds. Hardier plants are generated through plant breeding rather than genetic engineering.

While organic is fundamentally different from conventional because of the use of carbon based fertilizers compared with highly soluble synthetic based fertilizers and biological pest control instead of synthetic pesticides, organic farming and large-scale conventional farming are not entirely mutually exclusive. Many of the methods developed for organic agriculture have been borrowed by more conventional agriculture. For example, Integrated Pest Management is a multifaceted strategy that uses various organic methods of pest control whenever possible, but in conventional farming could include synthetic pesticides only as a last resort.

Crop Diversity

Crop diversity is a distinctive characteristic of organic farming. Conventional farming focuses on mass production of one crop in one location, a practice called monoculture. The science of agroecology has revealed the benefits of polyculture (multiple crops in the same space), which is often employed in organic farming. Planting a variety of vegetable crops supports a wider range of beneficial insects, soil microorganisms, and other factors that add up to overall farm health. Crop diversity helps environments thrive and protect species from going extinct.

Soil Management

Organic farming relies heavily on the natural breakdown of organic matter, using techniques like green manure and composting, to replace nutrients taken from the soil by previous crops. This biological process, driven by microorganisms such as mycorrhiza, allows the natural production of nutrients in the soil throughout the growing season, and has been referred to as *feeding the soil to feed the plant.* Organic farming uses a variety of methods to improve soil fertility, including crop rotation, cover cropping, reduced tillage, and application of compost. By reducing tillage, soil is not inverted and exposed to air; less carbon is lost to the atmosphere resulting in more soil organic carbon. This has an added benefit of carbon sequestration which can reduce green house gases and aid in reversing climate change.

Plants need nitrogen, phosphorus, and potassium, as well as micronutrients and symbiotic relationships with fungi and other organisms to flourish, but getting enough nitrogen, and particularly synchronization so that plants get enough nitrogen at the right time (when plants need it most), is a challenge for organic farmers. Crop rotation and green manure ("cover crops") help to provide nitrogen through legumes (more precisely, the *Fabaceae* family) which fix nitrogen from the atmosphere through symbiosis with rhizobial bacteria.

Intercropping, which is sometimes used for insect and disease control, can also increase soil nutrients, but the competition between the legume and the crop can be problematic and wider spacing between crop rows is required. Crop residues can be ploughed back into the soil, and different plants leave different amounts of nitrogen, potentially aiding synchronization. Organic farmers also use animal manure, certain processed fertilizers such as seed meal and various mineral powders such as rock phosphate and greensand, a naturally occurring form of potash which provides potassium. Together these methods help to control erosion. In some cases pH may need to be amended. Natural pH amendments include lime and sulfur, but in the U.S. some compounds such as iron sulfate, aluminum sulfate, magnesium sulfate, and soluble boron products are allowed in organic farming.

Mixed farms with both livestock and crops can operate as ley farms, whereby the land gathers fertility through growing nitrogen-fixing forage grasses such as white clover or alfalfa and grows cash crops or cereals when fertility is established. Farms without livestock ("stockless") may find it more difficult to maintain soil fertility, and may rely more on external inputs such as imported manure as well as grain legumes and green manures, although grain legumes may fix limited nitrogen because they are harvested. Horticultural farms growing fruits and vegetables which operate in protected conditions are often even more reliant upon external inputs.

Biological research on soil and soil organisms has proven beneficial to organic farming. Varieties of bacteria and fungi break down chemicals, plant matter and animal waste into productive soil nutrients. In turn, they produce benefits of healthier yields and more productive soil for future crops. Fields with less or no manure display significantly lower yields, due to decreased soil microbe community, providing a healthier, more arable soil system.

Weed Management

Organic weed management promotes weed suppression, rather than weed elimination, by enhancing crop competition and phytotoxic effects on weeds. Organic farmers integrate cultural, biological, mechanical, physical and chemical tactics to manage weeds without synthetic herbicides.

Organic standards require rotation of annual crops, meaning that a single crop cannot be grown in the same location without a different, intervening crop. Organic crop rotations frequently include weed-suppressive cover crops and crops with dissimilar life cycles to discourage weeds associated with a particular crop. Research is ongoing to develop organic methods to promote the growth of natural microorganisms that suppress the growth or germination of common weeds.

Other cultural practices used to enhance crop competitiveness and reduce weed pressure include selection of competitive crop varieties, high-density planting, tight row spacing, and late planting into warm soil to encourage rapid crop germination.

Mechanical and physical weed control practices used on organic farms can be broadly grouped as:

- Tillage - Turning the soil between crops to incorporate crop residues and soil amendments; remove existing weed growth and prepare a seedbed for planting; turning soil soil after seeding to kill weeds;
- Mowing and cutting - Removing top growth of weeds;
- Flame weeding and thermal weeding - Using heat to kill weeds; and
- Mulching - Blocking weed emergence with organic materials, plastic films, or landscape fabric.

Some critics, citing work published in 1997 by David Pimentel of Cornell University, which described an epidemic of topsoil erosion worldwide, have raised concerned that tillage contribute to the erosion epidemic. The FAO and other organisations have advocated a "no-till" approach to both conventional and organic farming, and point out in particular that crop rotation techniques used in organic farming are excellent no-till approaches. A study published in 2005 by Pimentel and colleagues confirmed that "Crop rotations and cover cropping (green manure) typical of organic agriculture reduce soil erosion, pest problems, and pesticide use." Some naturally sourced chemicals are allowed for herbicidal use. These include certain formulations of acetic acid (concentrated vinegar), corn gluten meal, and essential oils. A few selective bioherbicides based on fungal pathogens have also been developed. At this time, however, organic herbicides and bioherbicides play a minor role in the organic weed control toolbox.

Weeds can be controlled by grazing. For example, geese have been used successfully to weed a range of organic crops including cotton, strawberries, tobacco, and corn, reviving the practice of keeping cotton patch geese, common in the southern U.S. before the 1950s. Similarly, some rice farmers introduce ducks and fish to wet paddy fields to eat both weeds and insects.

Controlling Other Organisms

Organisms aside from weeds that cause problems on organic farms include arthropods (e.g., insects, mites), nematodes, fungi and bacteria. Organic practices include, but are not limited to:

- encouraging predatory beneficial insects to control pests by serving them nursery plants and/or an alternative habitat, usually in a form of a shelterbelt, hedgerow, or beetle bank;
- encouraging beneficial microorganisms;
- rotating crops to different locations from year to year to interrupt pest reproduction cycles;
- planting companion crops and pest regulating plants that discourage or divert pests;
- using row covers to protect crops during pest migration periods;
- using biologic pesticides and herbicides
- using no-till farming, and no-till farming techniques as false seedbeds
- using sanitation to remove pest habitat;
- Using insect traps to monitor and control insect populations.
- Using physical barriers, such as row covers

Examples of predatory beneficial insects include minute pirate bugs, big-eyed bugs, and to a lesser extent ladybugs (which tend to fly away), all of which eat a wide range of pests. Lacewings are also effective, but tend to fly away. Praying mantis tend to move more slowly and eat less heavily. Parasitoid wasps tend to be effective for their selected prey, but like all small insects can be less effective outdoors because the wind controls their movement. Predatory mites are effective for controlling other mites.

Naturally derived insecticides allowed for use on organic farms use include *Bacillus thuringiensis* (a bacterial toxin), pyrethrum (a chrysanthemum extract), spinosad (a bacterial metabolite), neem (a tree extract) and rotenone (a legume root extract). Fewer than 10% of organic farmers use these pesticides regularly; one survey found that only 5.3% of vegetable growers in California use rotenone while 1.7% use pyrethrum. These pesticides are not always more safe or environmentally friendly than synthetic pesticides and can cause harm. The main criterion for organic pesticides is that they are naturally derived, and some naturally derived substances have been controversial. Controversial natural pesticides include rotenone, copper, nicotine sulfate, and pyrethrums Rotenone and pyrethrum are particularly controversial because they work by attacking the nervous system, like most conventional insecticides. Rotenone is extremely toxic to fish and can induce symptoms resembling Parkinson's disease in mammals. Although pyrethrum (natural pyrethrins) is more effective against insects when used with piperonyl butoxide (which retards degradation of the pyrethrins), organic standards generally do not permit use of the latter substance.

Naturally derived fungicides allowed for use on organic farms include the bacteria *Bacillus subtilis* and *Bacillus pumilus*; and the fungus *Trichoderma harzianum*. These are mainly effective for diseases affecting roots. Compost tea contains a mix of beneficial microbes, which may attack or out-compete certain plant pathogens, but variability among formulations and preparation methods may contribute to inconsistent results or even dangerous growth of toxic microbes in compost teas.

Some naturally derived pesticides are not allowed for use on organic farms. These include nicotine sulfate, arsenic, and strychnine.

Synthetic pesticides allowed for use on organic farms include insecticidal soaps and horticultural oils for insect management; and Bordeaux mixture, copper hydroxide and sodium bicarbonate for managing fungi. Copper sulfate and Bordeaux mixture (copper sulfate plus lime), approved for organic use in various jurisdictions, can be more environmentally problematic than some synthetic fungicides dissallowed in organic farming Similar concerns apply to copper hydroxide.

Repeated application of copper sulfate or copper hydroxide as a fungicide may eventually result in copper accumulation to toxic levels in soil, and admonitions to avoid excessive accumulations of copper in soil appear in various organic standards and elsewhere. Environmental concerns for several kinds of biota arise at average rates of use of such substances for some crops. In the European Union, where replacement of copper-based fungicides in organic agriculture is a policy priority, research is seeking alternatives for organic production.

Figure: *Chloroxylon is used for Pest Management in Organic Rice Cultivation in Chhattisgarh, India*

Livestock

Raising livestock and poultry, for meat, dairy and eggs, is another traditional, farming activity that complements growing. Organic farms attempt to provide animals with "natural" living conditions and feed. While the USDA does not require any animal welfare requirements be met for a product to be marked as organic, this is a variance from older organic farming practices.

Also, horses and cattle used to be a basic farm feature that provided labour, for hauling and plowing, fertility, through recycling of manure, and fuel, in the form of food for farmers and other animals. While today, small growing operations often do not include livestock, domesticated animals are a desirable part of the organic farming equation, especially for true sustainability, the ability of a farm to function as a self-renewing unit.

Genetic Modification

A key characteristic of organic farming is the rejection of genetically engineered plants and animals. On October 19, 1998, participants at IFOAM's 12th Scientific Conference issued the Mar del Plata Declaration, where more than 600 delegates from over 60 countries voted unanimously to exclude the use of genetically modified organisms in food production and agriculture. Although opposition to the use of any transgenic technologies in organic farming is strong, agricultural researchers Luis

Herrera-Estrella and Ariel Alvarez-Morales continue to advocate integration of transgenic technologies into organic farming as the optimal means to sustainable agriculture, particularly in the developing world, as does author and scientist Pamela Ronald, who views this kind of biotechnology as being consistent with organic principles.

Although GMOs are excluded from organic farming, there is concern that the pollen from genetically modified crops is increasingly penetrating organic and heirloom seed stocks, making it difficult, if not impossible, to keep these genomes from entering the organic food supply. Differing regulations among countries limits the availability of GMOs to certain countries, as described in the article on regulation of the release of genetic modified organisms.

Standards

Standards regulate production methods and in some cases final output for organic agriculture. Standards may be voluntary or legislated. As early as the 1970s private associations certified organic producers. In the 1980s, governments began to produce organic production guidelines. In the 1990s, a trend toward legislated standards began, most notably with the 1991 EU-Eco-regulation developed for European Union, which set standards for 12 countries, and a 1993 UK program. The EU's program was followed by a Japanese program in 2001, and in 2002 the U.S. created the National Organic Program (NOP). As of 2007 over 60 countries regulate organic farming (IFOAM 2007:11). In 2005 IFOAM created the Principles of Organic Agriculture, an international guideline for certification criteria. Typically the agencies accredit certification groups rather than individual farms.

Organic production materials used in and foods are tested independently by the Organic Materials Review Institute.

Composting

Under USDA organic standards, manure must be subjected to proper thermophilic composting and allowed to reach a sterilizing temperature. If raw animal manure is used, 120 days must pass before the crop is harvested if the final product comes into direct contact with the soil. For products which do not come into direct contact with soil, 90 days must pass prior to harvest.

Economics

The economics of organic farming, a subfield of agricultural economics, encompasses the entire process and effects of organic farming in terms of human society, including social costs, opportunity costs, unintended consequences, information asymmetries, and economies of scale. Although the scope of economics is broad, agricultural economics tends to focus on maximizing yields and efficiency at the farm level. Economics takes an anthropocentric approach to the value of the natural world: biodiversity, for example, is considered beneficial only to the extent that it is valued by people and increases profits.

Some entities such as the European Union subsidize organic farming, in large part because these countries want to account for the externalities of reduced water use, reduced water contamination, reduced soil erosion, reduced carbon emissions, increased biodiversity, and assorted other benefits that result from organic farming.

Traditional organic farming is labour and knowledge-intensive whereas conventional farming is capital-intensive, requiring more energy and manufactured inputs.

Organic farmers in California have cited marketing as their greatest obstacle.

Geographic Producer Distribution

The markets for organic products are strongest in North America and Europe, which as of 2001 are estimated to have \$6 and \$8 billion respectively of the \$20 billion global market. As of 2007 Australasia has 39% of the total organic farmland, including Australia's 1,180,000 hectares (2,900,000 acres) but 97 percent of this land is sprawling rangeland (2007:35). US sales are 20x as much. Europe farms 23 percent of global organic farmland (6.9 million hectares), followed by Latin America with 19 percent (5.8 million hectares). Asia has 9.5 percent while North America has 7.2 percent. Africa has 3 percent.

Besides Australia, the countries with the most organic farmland are Argentina (3.1 million hectares), China (2.3 million hectares), and the United States (1.6 million hectares). Much of Argentina's organic farmland is pasture, like that of Australia (2007:42). Spain, Germany, Brazil (the world's largest agricultural exporter), Uruguay, and the UK follow the United States in the amount of organic land (2007:26).

In the European Union (EU25) 3.9% of the total utilised agricultural area was used for organic production in 2005. The countries with the highest proportion of organic land were Austria (11%) and Italy (8.4), followed by the Czech Republic and Greece (both 7.2%). The lowest figures were shown for Malta (0.1%), Poland (0.6%) and

Ireland (0.8%). In 2009, the proportion of organic land in the EU grew to 4.7%. The countries with highest share of agricultural land were Liechtenstein (26.9%), Austria (18.5%) and Sweden (12.6%). 16% of all farmers in Austria produced organically in 2010. By the same year the proportion of organic land increased to 20%.: In 2005 168,000 ha of land in Poland was under organic management. In 2010 100,000 ha of land were under organic management in Romania, representing 1% of the total utilised agricultural area. 70%-80% of the local organic production, amounting to 100 million Euros in 2010, is exported. The organic products market grew to 50 million Euros in 2010.

After the collapse of the Soviet Union in 1991, agricultural inputs that had previously been purchased from Eastern bloc countries were no longer available in Cuba, and many Cuban farms converted to organic methods out of necessity. Consequently, organic agriculture is a mainstream practice in Cuba, while it remains an alternative practice in most other countries. Although some products called organic in Cuba would not satisfy certification requirements in other countries (crops may be genetically modified, for example), Cuba exports organic citrus and citrus juices to EU markets that meet EU organic standards. Cuba's forced conversion to organic methods may position the country to be a global supplier of organic products.

Growth

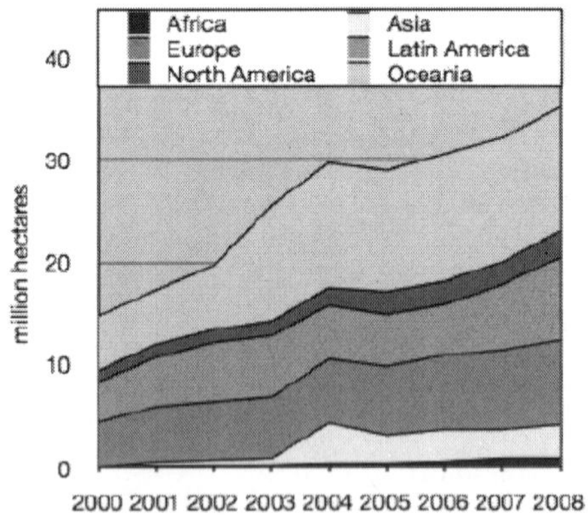

Figure: *Organic farmland by world region (2000-2008)*

As of 2001, the estimated market value of certified organic products was estimated to be \$20 billion. By 2002 this was \$23 billion and by 2007 more than \$46 billion.

In recent years both Europe (2007: 7.8 million hectares, European Union: 7.2 million hectares) and North America (2007: 2.2 million hectares) have experienced strong growth in organic farmland. In the EU it grew by 21% in the period 2005 to 2008. However, this growth has occurred under different conditions.

While the European Union has shifted agricultural subsidies to organic farmers due to perceived environmental benefits, the United States has not, continuing to subsidize some but not all traditional commercial crops, such as corn and sugar. As a result of this policy difference, as of 2008 4.1% percent of European Union farmland was organically managed compared to the 0.6 percent in the U.S.

IFOAM's most recent edition of *The World of Organic Agriculture: Statistics and Emerging Trends 2009* lists the countries which had the most hectares in 2007. The country with the most organic land is Australia with more than 12 million hectares, followed by Argentina, Brazil and the US. In total 32.2 million hectares were under organic management in 2007. For 1999 11 million hectares of organically managed land are reported.

As organic farming becomes a major commercial force in agriculture, it is likely to gain increasing impact on national agricultural policies and confront some of the scaling challenges faced by conventional agriculture.

Productivity

Studies comparing yields have had mixed results. A study published in 1990 made "two hundred and five comparisons ... of yields from organic and conventional farming systems..... Data from 26 crops and two animal products, in the form of the ratio of organic to conventional yields, were normally distributed with a mean of 0.91, a standard deviation of 0.24 and a modal value between 0.8 and 0.9. More than one-half of the comparisons of milk production and bean yields had ratios greater than 1.0, i.e. higher yields from organic than conventional systems. There was no evidence to show that the organic systems had any effect on year-to-year variability in yield, either climate-induced or caused by any transitional or conversion effects." The study also discussed procedural difficulties in comparing the productivity of organic with other farming systems.

A US survey published in 2001 analyzed 150 growing seasons of data on grain and soybean crops and concluded that organic yields were 95-100% of conventional yields.

A study spanning two decades was published in 2002 and found a 20% smaller yield from organic farms using 50% less fertilizer, 97%

less pesticide, and energy input was 34% to 53% lower. A 2003 study found that during drought years, organic farms can have yields 20-40% higher than conventional farms. Organic farms are more profitable in the drier states of the United States, likely due to their superior drought performance.

Organic farms survive hurricane damage much better, retaining 20 to 40% more topsoil and smaller economic losses at highly significant levels than their neighbours.

A study published in 2005 compared conventional cropping, organic animal-based cropping, and organic legume-based cropping on a test farm at the Rodale Institute over 22 years. The study found that "the crop yields for corn and soybeans were similar in the organic animal, organic legume, and conventional farming systems". It also found that "significantly less fossil energy was expended to produce corn in the Rodale Institute's organic animal and organic legume systems than in the conventional production system. There was little difference in energy input between the different treatments for producing soybeans. In the organic systems, synthetic fertilizers and pesticides were generally not used". As of 2013 the Rodale study was ongoing and a thirty year anniversary report was published by Rodale in 2012.

A 2007 study compiling research from 293 different comparisons into a single study to assess the overall efficiency of the two agricultural systems has concluded that "organic methods could produce enough food on a global per capita basis to sustain the current human population, and potentially an even larger population, without increasing the agricultural land base." The researchers also found that while in developed countries, organic systems on average produce 92% of the yield produced by conventional agriculture, organic systems produce 80% more than conventional farms in developing countries, because the materials needed for organic farming are more accessible than synthetic farming materials to farmers in some poor countries. This study was strongly contested by another study published in 2008 which stated, and was entitled, "Organic agriculture cannot feed the world" and said that the 2007 came up with "a major overestimation of the productivity of OA" "because data are misinterpreted and calculations accordingly are erroneous."

Another study published in 1999 from the Danish Environmental Protection Agency found that, area-for-area, organic farms of potatoes, sugar beet and seed grass produce as little as half the output of conventional farming. Michael Pollan, author of *The Omnivore's*

Dilemma, responds to this by pointing out that the average yield of world agriculture is substantially lower than modern sustainable farming yields. Bringing average world yields up to modern organic levels could increase the world's food supply by 50%.

Profitability

The decreased cost of synthetic fertilizer and pesticide inputs, along with the higher prices that consumers pay for organic produce, contribute to increased profits. Organic farms have been consistently found to be as or more profitable than conventional farms. Without the price premium, profitability is mixed. Organic production was more profitable in Wisconsin, given price premiums.

For markets and supermarkets organic food is profitable as well, and is generally even sold at significantly higher rates than non-organic food

However, when the buyer compares prices and buys consciously, organic food is not always more expensive for the buyer than non-organic food. For example, in 2000, Phillipe Renard made his restaurant to switch to use 85% organic food, without increasing the cost for the clients. In the documentary "Architects for Change", he also stated that since 2000, the cost of organic products has come down even more, and at present, it is no longer a problem to attain organic products at a price comparative to products of non-organic agriculture.

Energy Efficiency

A study of the sustainability of apple production systems showed that in comparing a conventional farming system to an organic method of farming, the organic system in this case is more energy efficient. A more comprehensive study compared efficiency of agriculture for products such as grain, roughage crops, and animal husbandry. While the study did not investigate specific additional requirements of arable land or numbers of farm labourers to produce total yields for organic farming vs.

conventional farming, leaving open the question of overall capacity of organic farming to meet current and future agricultural needs, it concluded that organic farming had a higher yield per unit of energy over multiple crops and for livestock. However, conventional farming had higher total yield. Conversely, another study noted that organic wheat and corn production was more energy efficient than conventional methods while organic apple and potato production was less energy efficient than conventional methods.

A study done with apple orchards in the state of Washington found that organic orchards found to be at least 7% more energy efficient.

Sales and Marketing

Most sales are concentrated in developed nations. These products are what economists call credence goods in that they rely on uncertain certification. Interest in organic products dropped between 2006 and 2008, and 42% of Americans polled don't trust organic produce. 69% of Americans claim to occasionally buy organic products, down from 73% in 2005. One theory was that consumers were substituting "local" produce for "organic" produce.

Distributors

In the United States, 75% of organic farms are smaller than 2.5 hectares. In California 2% of the farms account for over half of sales. Small farms join together in cooperatives such as Organic Valley, Inc. to market their goods more effectively.

Most small cooperative distributors have merged or were acquired by large multinationals such as General Mills, Heinz, ConAgra, Kellogg, and others. In 1982 there were 28 consumer cooperative distributors, but as of 2007 only 3 remained. This consolidation has raised concerns among consumers and journalists of potential fraud and degradation in standards. Most sell their organic products through subsidiaries, under other labels.

Organic foods also can be a niche in developing nations. It would provide more money and a better opportunity to compete internationally with the huge distributors. Organic prices are much more stable than conventional foods, and the small farms can still compete and have similar prices with the much larger farms that usually take all of the profits.

Farmers Markets

Price premiums are important for the profitability of small organic farmers. Farmers selling directly to consumers at farmers' markets have continued to achieve these higher returns. In the United States the number of farmers' markets tripled from 1,755 in 1994 to 5,274 in 2009.

Employment

In a survey of 1144 organic farms in Ireland and the United Kingdom, researchers found that organic farms employed more workers than conventional ones. This difference persisted when factors such as the size of each farm (organic farms are typically larger) are taken into account. The researchers concluded that there would be 19% more farming jobs in the UK, and 6% more in Ireland, if 20% of all farms became organic.

Capacity Building in Developing Countries

Organic agriculture can contribute to ecologically sustainable, socio-economic development, especially in poorer countries. The application of organic principles enables employment of local resources (e.g., local seed varieties, manure, etc.) and therefore cost-effectiveness. Local and international markets for organic products show tremendous growth prospects and offer creative producers and exporters excellent opportunities to improve their income and living conditions.

Organic agriculture is knowledge intensive. Globally, capacity building efforts are underway, including localized training material, to limited effect. As of 2007, the International Federation of Organic Agriculture Movements hosted more than 170 free manuals and 75 training opportunities online.

In 2008 the United Nations Environmental Programme (UNEP) and the United Nations Conference on Trade and Development (UNCTAD) stated that "organic agriculture can be more conducive to food security in Africa than most conventional production systems, and that it is more likely to be sustainable in the long-term" and that "yields had more than doubled where organic, or near-organic practices had been used" and that soil fertility and drought resistance improved.

Organic Agriculture and the Millennium Development Goals (MDGs)

The value of organic agriculture (OA) in the achievement of the MDGs particularly in poverty reduction efforts in the face of climate change can be shown in its contribution to both income and non-income aspects of the MDGs. A series of case studies conducted by the Asian Development Bank Institute (ADBI) in Tokyo showed that OA contributes to both income and non-income aspects of the MDGs in all the study areas in selected Asian countries. OA's outcomes on MDGs include contributions to the alleviation of poverty by way of higher incomes, improved farmers' health owing to less chemical exposure, integration of sustainable principles into rural development policies, improvement of access to safe water and sanitation, and expansion of global partnership for development.

A related ADBI study on OA estimates costs of OA programs and sets them in the context of the costs of attaining the MDGs. The results show considerable variation across the case studies, suggesting that there is no clear structure to the costs of adopting OA. Costs depend on the efficiency with which the OA adoption programs are run. The lowest cost programs were more than ten times less expensive than the highest cost ones. A further analysis of the gains resulting from OA adoption reveals that the costs per person taken out of poverty was much lower than the estimates of the World Bank, based on income growth in general or based on the detailed costs of meeting some of the more quantifiable MDGs (e.g., education, health, and environment).

Externalities

Agriculture imposes negative externalities (uncompensated costs) upon society through land and other resource use, biodiversity loss, erosion, pesticides, nutrient runoff, water usage, subsidy payments and assorted other problems. Positive externalities include self-reliance, entrepreneurship, respect for nature, and air quality. Organic methods reduce some of these costs. In 2000 uncompensated costs for 1996 reached 2,343 million British pounds or 208 pounds per hectare. A study of practices in the USA published in 2005 concluded that cropland costs the economy approximately 5 to 16 billion dollars ($30 to $96 per hectare), while livestock production costs 714 million dollars. Both studies recommended reducing externalities. The 2000 review included reported pesticide poisonings but did not include speculative chronic health effects of pesticides, and the 2004 review relied on a 1992 estimate of the total impact of pesticides.

It has been proposed that organic agriculture can reduce the level of some negative externalities from (conventional) agriculture. Whether the benefits are private or public depends upon the division of property rights.

Several surveys and studies have attempted to examine and compare conventional and organic systems of farming and have found that organic techniques, while not without harm, are less damaging than conventional ones because they reduce levels of biodiversity less than conventional systems do and use less energy and produce less waste when calculated per unit area.

A 2003 investigation by the Department for Environment Food and Rural Affairs in the UK found, similar to other reports, that organic farming "can produce positive environmental benefits", but that some of the benefits were decreased or lost when comparisons are made on "the basis of unit production rather than area".

The situation was shown dramatically in a comparison of a modern dairy farm in Wisconsin with one in New Zealand in which the animals grazed extensively. Using total farm emissions per kg milk produced as a parameter, the researchers showed that production of methane from belching was higher in the New Zealand farm, while carbon dioxide production was higher in the Wisconsin farm. Output of nitrous oxide, a gas with an estimated global warming potential 310 times that of carbon dioxide was also higher in the New Zealand farm. Methane from manure handling was similar in the two types of farm. The explanation for the finding relates to the different diets used on these farms, being based more completely on forage (and hence more fibrous) in New Zealand and containing less concentrate than in Wisconsin. Fibrous diets promote a higher proportion of acetate in the gut of ruminant animals, resulting in a higher production of methane that has to be released by belching. When cattle are given a diet containing some concentrates (such as corn and soybean meal) in addition to grass and silage, the pattern of ruminal fermentation alters from acetate to mainly propionate. As a result methane production is reduced. Capper et al. compared the environmental impact of US dairy production in 1944 and 2007. They calculated that the carbon "footprint" per billion kg of milk produced in 2007 was 37 percent that of equivalent milk production in 1944.

Pesticides

Unlike conventional farms, most organic farms largely avoid synthetic pesticides Some pesticides damage the environment or with direct exposure, human health. Children may be more at risk than adults from direct exposure, as the toxicity of pesticides is frequently different in children and adults.

The five main pesticides used in organic farming are Bt (a bacterial toxin), pyrethrin, rotenone, copper and sulphur. "Fewer than 10% of organic farmers use botanical insecticides on a regular basis, 12% use sulfur, and 7% use copper-based compounds." Reduction and elimination of chemical pesticide use is technically challenging. Organic pesticides often complement other pest control strategies.

Ecological concerns primarily focus around pesticide use, as 16% of the world's pesticides are used in the production of cotton.

Runoff is one of the most damaging effects of pesticide use. The USDA Natural Resources Conservation Service tracks the environmental effects of water contamination and concluded, "the Nation's pesticide policies during the last twenty six years have succeeded in reducing overall environmental risk, in spite of slight increases in area planted and weight of pesticides applied. Nevertheless, there are still areas of the country where there is no evidence of progress, and areas where risk levels for protection of drinking water, fish, algae and crustaceans remain high".

Food Quality and Safety

The weight of the available scientific evidence has not shown a consistent and significant difference between organic and more conventionally grown food in terms of safety, or nutritional value. In 2009 a review of all the relevant research comparing organic to conventionally grown foods was carried out by the United Kingdom's Food Standards Agency which concluded:

No evidence of a difference in content of nutrients and other substances between organically and conventionally produced crops and livestock products was detected for the majority of nutrients assessed in this review suggesting that organically and conventionally produced crops and livestock products are broadly comparable in their nutrient content... There is no good evidence that increased dietary intake, of the nutrients identified in this review to be present in larger amounts in organically than in conventionally produced crops and livestock products, would be of benefit to individuals consuming a normal varied diet, and it is therefore unlikely that these differences in nutrient content are relevant to consumer health.

A 2009 review of potential health effects conducted for the UK Food Standards Agency analysed eleven articles, concluding, "because of the limited and highly variable data available, and concerns over the reliability of some reported findings, there is currently no evidence of a health benefit from consuming organic compared to conventionally produced foodstuffs. It should be noted that this conclusion relates to the evidence base currently available on the nutrient content of foodstuffs, which contains limitations in the design and in the comparability of studies."

Individual studies have considered a variety of possible impacts, including pesticide residues. Pesticide residues present a second channel for health effects. Comments include, "Organic fruits and vegetables can be expected to contain fewer agrochemical residues than conventionally grown alternatives; yet, "the significance of this difference is questionable".

Nitrate concentrations may be less, but the health impact of nitrates is debated. Lack of data has limited research on the health effects of natural plant pesticides and bacterial pathogens.

The higher cost of organic food (ranging from 45 to 200%) could inhibit consumption of the recommended 5 servings per day of vegetables and fruits, which improve health and reduce cancer regardless of their source.

Soil Conservation

Supporters claim that organically managed soil has a higher quality and higher water retention. This may help increase yields for organic farms in drought years. Organic farming can build up soil organic matter better than conventional no-till farming, which suggests long-term yield benefits from organic farming. An 18-year study of organic methods on nutrient-depleted soil, concluded that conventional methods were superior for soil fertility and yield for nutrient-depleted soils in cold-temperate climates, arguing that much of the benefits from organic farming are derived from imported materials which could not be regarded as "self-sustaining".

In Dirt: *The Erosion of Civilizations*, geomorphologist David Montgomery outlines a coming crisis from soil erosion. Agriculture relies on roughly one metre of topsoil, and that is being depleted ten times faster than it is being replaced. No-till farming, which some claim depends upon pesticides, is one way to minimize erosion. However, a recent study by the USDA's Agricultural Research Service has found that manure applications in tilled organic farming are better at building up the soil than no-till.

Climate Change

Organic agriculture emphasizes closed nutrient cycles, biodiversity, and effective soil management providing the capacity to mitigate and even reverse the effects of climate change. Organic agriculture can decrease fossil fuel emissions and, like any well managed agricultural system, sequesters carbon in the soil. The elimination of synthetic nitrogen in organic systems decreases fossil fuel consumption by 33 percent and carbon sequestration takes CO_2 out of the atmosphere by putting it in the soil in the form of organic matter which is often lost in conventionally managed soils. Carbon sequestration occurs at especially high levels in organic no-till managed soil.

Agriculture has been undervalued and underestimated as a means to combat global climate change. Soil carbon data show that regenerative organic agricultural practices are among the most effective strategies for mitigating CO_2 emissions.

However, critics of organic farming methods believe that the increased land needed to farm organic food could potentially destroy the rainforests and wipe out many ecosystems.

Nutrient Leaching

Excess nutrients in lakes, rivers, and groundwater can cause algal blooms, eutrophication, and subsequent dead zones. In addition, nitrates are harmful to aquatic organisms by themselves. The main contributor to this pollution is nitrate fertilizers whose use is expected to "double or almost triple by 2050". Organically fertilizing fields "significantly [reduces] harmful nitrate leaching" over conventionally fertilized fields: "annual nitrate leaching was 4.4-5.6 times higher in conventional plots than organic plots".

The large dead zone in the Gulf of Mexico is caused in large part by agricultural runoff: a combination of fertilizer and livestock manure. Over half of the nitrogen released into the Gulf comes from agriculture. This increases costs for fishermen, as they must travel far from the coast to find fish. Nitrogen leaching into the Danube River was substantially lower among organic farms. The resulting externalities could be neutralized by charging 1 euro per kg of released nitrogen.

Agricultural runoff and algae blooms are strongly linked in California.

Biodiversity

A wide range of organisms benefit from organic farming, but it is unclear whether organic methods confer greater benefits than conventional integrated agri-environmental programs. Nearly all non-crop, naturally occurring species observed in comparative farm land practice studies show a preference for organic farming both by abundance and diversity. An average of 30% more species inhabit organic farms. Birds, butterflies, soil microbes, beetles, earthworms, spiders, vegetation, and mammals are particularly affected. Lack of herbicides and pesticides improve biodiversity fitness and population density. Many weed species attract beneficial insects that improve soil qualities and forage on weed pests. Soil-bound organisms often benefit because of increased bacteria populations due to natural fertilizer such as manure, while experiencing reduced intake of herbicides and pesticides. Increased biodiversity, especially from beneficial soil microbes and mycorrhizae have been proposed as an explanation for the high yields experienced by some organic plots, especially in light of the differences seen in a 21-year comparison of organic and control fields.

Biodiversity from organic farming provides capital to humans. Species found in organic farms enhance sustainability by reducing human input (e.g., fertilizers, pesticides).

Critical Analysis

Norman Borlaug (father of the "Green Revolution" and a Nobel Peace Prize laureate), Prof A. Trewavas and other critics contested the notion that organic agricultural systems are more friendly to the environment and more sustainable than conventional farming systems. Borlaug asserts that organic farming practices can at most feed 4 billion people, after expanding cropland dramatically and destroying ecosystems in the process. Borlaug and his co-authors advocated using organic matter in addition to inorganic fertilizers in soil fertility management, but opposed advocating only organic agriculture for the developing world. The Danish Environmental Protection Agency estimated that phasing out all pesticides would result in an overall yield reduction of about 25%. Environmental and health effects were assumed but hard to assess.

One study claims that organic agriculture could feed the entire global population, somewhat more than 6 billion people. It states that organic farms have lower yields than their conventional counterparts in developed countries (92%) but higher than their low-intensity counterparts in developing countries (180%), attributing this to lower adoption of fertilizers and pesticides in the developing world compared to the intensive farming of the developed world. However, concerns have been expressed about that study's selection, characterization and interpretation of data, and its assumptions and analytical methods, casting doubt on several of its conclusions. The Centres for Disease Control repudiated a claim by Dennis Avery of the Hudson Institute, that the risk of E. coli infection was eight times higher when eating organic food. (Avery had cited CDC as a source.) Avery had included problems stemming from non-organic unpasteurized juice in his calculations. Epidemiologists traced the 2011 E. coli O104:H4 outbreak - which caused over 3,900 cases and 52 deaths - to an organic farm in Bienenbüttel in Germany.

A long-term field study comparing organic/conventional agriculture carried out over 21 years in Switzerland concluded that "Crop yields of the organic systems averaged over 21 experimental years at 80%

of the conventional ones. The fertilizer input, however, was 34 – 51% lower, indicating an efficient production. The organic farming systems used 20 – 56% less energy to produce a crop unit and per land area this difference was 36 – 53%. In spite of the considerably lower pesticide input the quality of organic products was hardly discernible from conventional analytically and even came off better in food preference trials and picture creating methods"

Genetic Approach in Vegetable Farming

Genetically modified foods (GM foods, or biotech foods) are foods produced from genetically modified organisms (GMOs), specifically, genetically modified crops. GMOs have had specific changes introduced into their DNA by genetic engineering techniques. These techniques are much more precise than mutagenesis (mutation breeding) where an organism is exposed to radiation or chemicals to create a non-specific but stable change. Other techniques by which humans modify food organisms include selective breeding and somaclonal variation.

Commercial sale of genetically modified foods began in 1994, when Calgene first marketed its Flavr Savr delayed ripening tomato. Typically, genetically modified foods are transgenic plant products: soybean, corn, canola, and cotton seed oil. These may have been engineered for faster growth, resistance to pathogens, production of extra nutrients, or any other beneficial purpose. GM livestock have also been experimentally developed, although as of July 2010 none are currently on the market.

There is broad scientific consensus that food on the market derived from GM crops poses no greater risk to human health than conventional food. However, critics have objected to GM foods on several grounds, including safety issues, ecological concerns, and economic concerns raised by the fact that GM plants (and potentially animals) that are food sources are subject to intellectual property law.

Method of Production

Genetically engineered plants are generated in a laboratory by altering their genetic makeup and are tested in the laboratory for desired qualities. This is usually done by adding one or more genes to a plant's genome using genetic engineering techniques. Most genetically modified plants are generated by the biolistic method (particle gun) or by *Agrobacterium tumefaciens* mediated transformation. Once satisfactory plants are produced, sufficient seeds are gathered, and the companies producing the seed need to apply for regulatory approval to field-test the seeds. If these field tests are successful, the company must seek regulatory approval for the crop to be marketed. Once that approval is obtained, the seeds are mass-produced, and sold to farmers. The farmers produce genetically modified crops, which also contain the inserted gene and its protein product. The farmers then sell their crops as commodities into the food supply market, in countries where such sales are permitted.

History

Scientists first discovered that DNA can transfer between organisms in 1946. The first genetically modified plant was produced in 1983, using an antibiotic-resistant tobacco plant. In 1994, the transgenic Flavr Savr tomato was approved by the FDA for marketing in the US - the modification allowed the tomato to delay ripening after picking.

In the early 1990s, recombinant chymosin was approved for use in several countries, replacing rennet in cheese-making. In the US in 1995, the following transgenic crops received marketing approval: canola with modified oil composition (Calgene), *Bacillus thuringiensis* (Bt) corn/maize (Ciba-Geigy), cotton resistant to the herbicide bromoxynil (Calgene), Bt cotton (Monsanto), Bt potatoes (Monsanto), soybeans resistant to the herbicide glyphosate (Monsanto), virus-resistant squash (Monsanto-Asgrow), and additional delayed ripening tomatoes (DNAP, Zeneca/Peto, and Monsanto).

In 2000, with the creation of golden rice, scientists genetically modified food to increase its nutrient value for the first time. As of 2011, the U.S. leads a list of multiple countries in the production of GM crops, and 25 GM crops had received regulatory approval to be grown commercially. As of 2013, roughly 85% of corn, 91% of soybeans, and 88% of cotton produced in the United States are genetically modified.

Foods with Protein or DNA Remaining from GMOs

Currently, there are several GM crops that are food sources. In some cases, the product is directly consumed as food, but In most cases, crops that have been genetically modified are sold as commodities, which are further processed into food ingredients.

Fruits and Vegetables

Papaya has been genetically modified to resist the ringspot virus. 'SunUp' is a transgenic red-fleshed Sunset cultivar that is homozygous for the coat protein gene of PRV; 'Rainbow' is a yellow-fleshed F1 hybrid developed by crossing 'SunUp' and nontransgenic yellow-fleshed 'Kapoho'. The New York Times stated that "in the early 1990s, Hawaii's papaya industry was facing disaster because of the deadly papaya ringspot virus. Its single-handed savior was a breed engineered to be resistant to the virus. Without it, the state's papaya industry would have collapsed. Today, 80% of Hawaiian papaya is genetically engineered, and there is still no conventional or organic method to control ringspot virus."

The New Leaf potato, brought to market by Monsanto in the late 1990s, was developed for the fast food market, but was withdrawn from the market in 2001 after fast food retailers did not pick it up and food processors ran into export problems.

There are currently no transgenic potatoes marketed for human consumption. However, in October 2011 BASF requested cultivation and marketing approval as a feed and food from the EFSA for its Fortuna potato, which was made resistant to late blight by adding two resistance genes, blb1 and blb2, which originate from the Mexican wild potato Solanum bulbocastanum.

As of 2005, about 13% of the zucchini grown in the US was genetically modified to resist three viruses; the zucchini is also grown in Canada. As of 2012, an apple that has been genetically modified to resist browning, known as the Nonbrowning Arctic apple produced by Okanagan Speciality Fruits, is awaiting regulatory approval in the US and Canada. A gene in the fruit has been modified such that the apple produces less polyphenol oxidase, a chemical that manifests the browning.

Genetically engineered plants are generated in a laboratory by altering their genetic makeup. This is usually done by adding one or more genes to a plant's genome using genetic engineering techniques. Most genetically modified plants are generated by the biolistic method (particle gun) or by *Agrobacterium tumefaciens* mediated transformation. Plant scientists, backed by results of modern comprehensive profiling of crop composition, point out that crops modified using GM techniques are less likely to have unintended changes than are conventionally bred crops.

In research tobacco and *Arabidopsis thaliana* are the most genetically modified plants, due to well developed transformation methods, easy propagation and well studied genomes. They serve as model organisms for other plant species. In the biolistic method, DNA is bound to tiny particles of gold or tungsten which are subsequently shot into plant tissue or single plant cells under high pressure. The accelerated particles penetrate both the cell wall and membranes. The DNA separates from the metal and is integrated into plant genome inside the nucleus. This method has been applied successfully for many cultivated crops, especially monocots like wheat or maize, for which transformation using *Agrobacterium tumefaciens* has been less successful. The major disadvantage of this procedure is that serious damage can be done to the cellular tissue.

Agrobacteria are natural plant parasites, and their natural ability to transfer genes provides another method for the development of genetically engineered plants. To create a suitable environment for themselves, these Agrobacteria insert their genes into plant hosts, resulting in a proliferation of plant cells near the soil level (crown gall). The genetic information for tumour growth is encoded on a mobile, circular DNA fragment (plasmid). When Agrobacterium infects a plant, it transfers this T-DNA to a random site in the plant genome. When used in genetic engineering the bacterial T-DNA is removed from the bacterial plasmid and replaced with the desired foreign gene. The bacterium is a vector, enabling transportation of foreign genes into plants. This method works especially well for dicotyledonous plants like potatoes, tomatoes, and tobacco. Agrobacteria infection is less successful in crops like wheat and maize.

Introducing new genes into plants requires a promoter specific to the area where the gene is to be expressed. For instance, if we want the gene to be expressed only in rice grains and not in leaves, then an endosperm-specific promoter would be used. The codons of the gene must also be optimized for the organism due to codon usage bias. The transgenic gene products should also be able to be denatured by heat so that they are destroyed during cooking.

Flavr Savr Tomato

The first commercialised genetically modified plants (Flavr Savr tomatoes) used RNAi technology, where the inserted DNA matched an endogenous gene already in the plant. When the inserted gene is expressed it can repress the translation of the endogenous protein. Host delivered RNAi systems are being developed, where the plant will express RNA that will interfere with insects, nematodes and other parasites' protein synthesis. This may provide a novel way of protecting plants from pests.

Glyphosate Resistance

One of the most famous kinds of GM crops are "Roundup Ready", or glyphosate-resistant. Glyphosate, (the active ingredient in Roundup) kills plants by interfering with the shikimate pathway in plants, which is essential for the synthesis of the aromatic amino acids phenylalanine, tyrosine and tryptophan. More specifically, glyphosate inhibits the enzyme 5-enolpyruvylshikimate-3-phosphate synthase (EPSPS).

The shikimate pathway is not present in animals, which instead obtain aromatic amino acids from their diet.

Some micro-organisms have a version of EPSPS that is resistant to glyphosate inhibition. One of these was isolated from an *Agrobacterium* strain CP4 (CP4 EPSPS) that was resistant to glyphosate. This CP4 EPSPS gene was cloned and transfected into soybeans. The CP4 EPSPS gene was engineered for plant expression by fusing the 5' end of the gene to a chloroplast transit peptide derived from the petunia EPSPS. This transit peptide was used because it had shown previously an ability to deliver bacterial EPSPS to the chloroplasts of other plants. The plasmid used to move the gene into soybeans was PV-GMGTO4. It contained three bacterial genes, two CP4 EPSPS genes, and a gene encoding beta-glucuronidase (GUS) from *Escherichia coli* as a marker. The DNA was injected into the soybeans using the particle acceleration method. Soybean cultivar A54O3 was used for the transformation. The expression of the GUS gene was used as the initial evidence of transformation. GUS expression was detected by a staining method in which the GUS enzyme converts a substrate into a blue precipitate. Those plants that showed GUS expression were then taken and sprayed with glyphosate, and their tolerance was tested over many generations.

Types of Genetic Engineering

Transgenic plants have genes inserted into them that are derived from another species. The inserted genes can come from species within the same kingdom (plant to plant) or between kingdoms (for example, bacteria to plant). In many cases the inserted DNA has to be modified slightly in order to correctly and efficiently express in the host organism. Transgenic plants are used to express proteins like the cry toxins from *Bacillus thuringiensis*, herbicide resistant genes and antigens for vaccinations Transgenic carrots have been used to produce the drug Taliglucerase alfa which is used to treat Gaucher's disease. In the laboratory, transgenic plants have been modified to increase their photosynthesis (currently about 2% at most plants to the theoretic potential of 9–10%. This is possible by changing the rubisco enzyme (i.e. changing C3 plants into C4 plants), by placing the rubisco in a carboxysome, by adding CO_2 pumps in the cell wall, by changing the leaf form/size. Plants have been engineered to exhibit bioluminescence which might one day be a sustainable alternative to electric lighting. Still other transgenic plants have been modified to fix ambient nitrogen in the plant.

Cisgenic plants are made using genes found within the same species or a closely related one, where conventional plant breeding can occur. Some breeders and scientists argue that cisgenic modification is useful for plants that are difficult to crossbreed by conventional means (such as potatoes), and that plants in the cisgenic category should not require the same level of legal regulation as other genetically modified organisms.

Business of GM Crops

The global value of biotech seed alone was US$13.2 billion in 2011, with the end product of commercial grain from biotech maize, soybean grain and cotton valued at approximately US$160 billion or more per year.

Players in agriculture business markets include seed companies, agrochemical companies, distributors, farmers, grain elevators, and universities that develop new crops and whose agricultural extensions advise farmers on best practices.

The largest share of the GMO crops planted globally are from seed created by the United States firm Monsanto. In 2007, Monsanto's trait technologies were planted on 246 million acres (1,000,000 km^2) throughout the world, a growth of 13 percent from 2006. However, patents on the first Monsanto products to enter the marketplace will begin to expire in 2014, democratizing Monsanto products. Syngenta, DuPont (especially via its Pioneer Hi-Bred subsidiary, and Bayer CropScience are also major players in the US and Europe. In addition, a 2007 report from the European Joint Research Commission predicts that by 2015, more than 40 per cent of new GM plants entering the global marketplace will have been developed in Asia.

In the corn market, Monsanto's triple-stack corn—which combines Roundup Ready 2-weed control technology with YieldGard (Bt) Corn Borer and YieldGard Rootworm insect control—is the market leader in the United States. U.S. corn farmers planted more than 32 million acres (130,000 km^2) of triple-stack corn in 2008, and it is estimated the product could be planted on 56 million acres (230,000 km^2) in 2014–2015. In the cotton market, Bollgard II with Roundup Ready

Flex was planted on approximately 5 million acres (20,000 km^2) of U.S. cotton in 2008.

According to the International Service for the Acquisition of Agri-Biotech Applications (ISAAA), in 2010 approximately 15 million farmers grew biotech crops in 29 countries. Over 90% of the farmers were resource-poor in developing countries. 6.5 million farmers in China and 6.3 million small farmers in India grew biotech crops (mostly Bacillus thuringiensis cotton). The Philippines, South Africa (biotech cotton, maize, and soybeans often grown by subsistence women farmers) and another twelve developing countries also grew biotech crops in 2009. 10 million more small and resource-poor farmers may have been secondary beneficiaries of Bt cotton in China.

According to a review published in 2012 and based on data from the late 1990s and early 2000s, much of the GM crop grown each year is used for livestock feed, and increased demand for meat will lead to increased demand for GM crops with which to feed them. Feed grain usage as a percentage of total crop production is 70% for corn and more than 90% of oil seed meals such as soybeans. About 65 million metric tons of GM corn grains and about 70 million metric tons of soybean meals derived from GM soybean are fed to livestock each year.

Uses, Actual and Proposed

GM crops grown today, or under experimental development, have been modified with traits intended to provide benefit to farmers, consumers, or industry. These traits include improved shelf life, disease resistance, stress resistance, herbicide resistance, pest resistance, production of useful goods such as biofuel or drugs, and ability to absorb toxins, for use in bioremediation of pollution. Due to high regulatory and research costs, the majority of genetically modified crops in agriculture consist of commodity crops, such as soybean, maize, cotton and rapeseed. Recently, some research and development has been targeted to enhancement of crops that are locally important in developing countries, such as insect-resistant cowpea for Africa and insect-resistant brinjal (eggplant) for India.

Improved Shelf Life

The first genetically modified crop approved for sale in the U.S. was the *FlavrSavr* tomato, which had a longer shelf life. It is no longer on the market. As of 2012, an apple that has been genetically modified to resist browning, known as the Nonbrowning Arctic apple produced by Okanagan Speciality Fruits, is awaiting regulatory approval in the US and Canada. A gene in the fruit has been modified such that the apple produces less polyphenol oxidase, a chemical that manifests the browning.

Improved Nutrition

The GM oilseed crops on the market today offer improved oil profiles for processing or healthier edible oils. The GM crops in development offer a wider array of environmental and consumer benefits such as nutritional enhancement and drought and stress tolerance. GM plants are being developed by both private companies and public research institutions such as CIMMYT, the International Maize and Wheat Improvement Centre. Other examples include a genetically modified cassava with lower cyanogen glucosides and enhanced with protein and other nutrients, while golden rice, developed by the International Rice Research Institute (IRRI), has been discussed as a possible cure for Vitamin A deficiency. An international group of academics has generated a vitamin-enriched corn derived from South African white corn variety M37W with 169x increase in beta carotene, 6x the vitamin C and 2x folate – it is not in production anywhere, but proves that this can be done.

Stress Resistance

Plants engineered to tolerate non-biological stresses like drought, frost, high Soil salinity, and nitrogen starvation or with increased nutritional value (*e.g.* Golden rice) were in development in 2011.

Herbicide Resistance

Tobacco plants have been engineered to be resistant to the herbicide bromoxynil. And many crops have created that are resistant to the herbicide glyphosate. As weeds have grown resistant to glyphosate and other herbicides used in concert with resistant GM crops, companies are developing crops engineered to become resistant to multiple herbicides to allow farmers to use a mixed group of two, three, or four different chemicals.

Pathogen Resistance – Insects or Viruses

Tobacco, corn, rice and many other crops, have been generated that express genes encoding for insecticidal proteins from Bacillus thuringiensis (Bt). Papaya, potatoes, and squash have been engineered to resist viral pathogens, such as cucumber mosaic virus which, despite its name, infects a wide variety of plants.

Production of Biofuels

Algae, both hybrid and GM, is under development by several companies for the production of biofuels. Jatropha has also been modified to improve its qualities for fuel product. Swiss-based Syngenta has received USDA approval to market a maize seed trademarked Enogen, which has been genetically modified to convert its own starch to sugar to speed the process of making ethanol for biofuel. In 2013, the Flemish Institute for Biotechnology was investigating poplar trees genetically engineered to contain less lignin so that they would be more suitable for conversion into biofuels. Lignin is the critical limiting factor when using wood to make bio-ethanol because lignin limits the accessibility of cellulose microfibrils to depolymerization by enzymes.

Production of Useful By-Products

Drugs: Bananas have been developed, but are not in production, that produce human vaccines against infectious diseases such as Hepatitis B. Tobacco plants have been developed and studied, but are not in production, that can produce therapeutic antibodies.

Materials

Several companies and labs are working on engineering plants that can be used to make bioplastics. Potatoes that produce more industrially useful starches have been developed as well.

Bioremediation

Scientists at the University of York developed a weed (*Arabidopsis thaliana*) that contains genes from bacteria that can clean up TNT and RDX-explosive contaminants from the soil: It was hoped that this weed would eliminate this pollution. 16 million hectares in the USA (1.5% of the total surface) are estimated to be contaminated with TNT and RDX. However the weed *Arabidopsis thaliana* was not tough enough to withstand the environment on military test grounds and research is continuing with the University of Washington to develop a tougher native grass. Genetically modified plants have also been used for bioremediation of contaminated soils. Mercury, selenium and organic pollutants such as polychlorinated biphenyls (PCBs), TNT and RDX explosive contaminants have been removed from soils by transgenic plants containing genes for bacterial enzymes.

Cover Crop

A cover crop is a crop planted primarily to manage soil fertility, soil quality, water, weeds, pests, diseases, biodiversity and wildlife in an *agroecosystem* (Lu *et al.* 2000), an ecological system managed and largely shaped by humans across a range of intensities to produce food, feed, or fiber.

Cover crops are of interest in sustainable agriculture as many of them improve the sustainability of agroecosystem attributes and may also indirectly improve qualities of neighbouring natural ecosystems. Farmers choose to grow and manage specific cover crop types based on their own needs and goals, influenced by the biological, environmental, social, cultural, and economic factors of the food system in which farmers operate (Snapp *et al.* 2005).

Soil Fertility Management

One of the primary uses of cover crops is to increase soil fertility. These types of cover crops are referred to as "green manure." They are used to manage a range of soil macronutrients and micronutrients. Of the various nutrients, the impact that cover crops have on nitrogen management has received the most attention from researchers and farmers, because nitrogen is often the most limiting nutrient in crop production. Often, green manure crops are grown for a specific period, and then plowed under before reaching full maturity in order to improve soil fertility and quality.

Green manure crops are commonly leguminous, meaning they are part of the Fabaceae (pea) family. This family is unique in that all of the species in it set pods, such as bean, lentil, lupins and Alfalfa. Leguminous cover crops are typically high in nitrogen and can often provide the required quantity of nitrogen for crop production. In conventional farming, this nitrogen is typically applied in chemical fertilizer form. This quality of cover crops is called fertilizer replacement value (Thiessen-Martens *et al.* 2005).

Another quality unique to leguminous cover crops is that they form symbiotic relationships with the rhizobial bacteria that reside in legume root nodules. Lupins is nodulated by the soil microorganism *Bradyrhizobium* sp. (Lupinus). Bradyrhizobia are encountered as microsymbionts in other leguminous crops (*Argyrolobium, Lotus, Ornithopus, Acacia, Lupinus*) of Mediterranean origin. These bacteria convert biologically unavailable atmospheric nitrogen gas (N_2) to biologically available ammonium ($NH+_4$) through the process of biological nitrogen fixation.

Prior to the advent of the Haber-Bosch process, an energy-intensive method developed to carry out industrial nitrogen fixation and create chemical nitrogen fertilizer, most nitrogen introduced to ecosystems arose through biological nitrogen fixation (Galloway *et al.* 1995). Some scientists believe that widespread biological nitrogen fixation, achieved mainly through the use of cover crops, is the only alternative to industrial nitrogen fixation in the effort to maintain or increase future food production levels (Bohlool *et al.* 1992, Peoples and Craswell 1992, Giller and Cadisch 1995). Industrial nitrogen fixation has been criticized as an unsustainable source of nitrogen for food production due to its reliance on fossil fuel energy and the environmental impacts associated with chemical nitrogen fertilizer use in agriculture (Jensen and Hauggaard-Nielsen 2003). Such widespread environmental impacts include nitrogen fertilizer losses into waterways, which can lead to eutrophication (nutrient loading) and ensuing hypoxia (oxygen depletion) of large bodies of water.

An example of this lies in the Mississippi Valley Basin, where years of fertilizer nitrogen loading into the watershed from agricultural production have resulted in a hypoxic "dead zone" off the Gulf of Mexico the size of New Jersey (Rabalais *et al.* 2002). The ecological complexity of marine life in this zone has been diminishing as a consequence (CENR 2000).

As well as bringing nitrogen into agroecosystems through biological nitrogen fixation, types of cover crops known as "catch crops" are used to retain and recycle soil nitrogen already present. The catch crops take up surplus nitrogen remaining from fertilization of the previous crop, preventing it from being lost through leaching (Morgan *et al.* 1942), or gaseous denitrification or volatilization (Thorup-Kristensen *et al.* 2003). Catch crops are typically fast-growing annual cereal species adapted to scavenge available nitrogen efficiently from the soil (Ditsch and Alley 1991). The nitrogen tied up in catch crop biomass is released back into the soil once the catch crop is incorporated as a green manure or otherwise begins to decompose.

An example of green manure use comes from Nigeria, where the cover crop *Mucuna pruriens* (velvet bean) has been found to increase the availability of phosphorus in soil after a farmer applies rock phosphate (Vanlauwe *et al.* 2000).

Soil Quality Management

Cover crops can also improve soil quality by increasing soil organic matter levels through the input of cover crop biomass over time. Increased soil organic matter enhances soil structure, as well as the water and nutrient holding and buffering capacity of soil (Patrick *et al.* 1957). It can also lead to increased soil carbon sequestration, which has been promoted as a strategy to help offset the rise in atmospheric carbon dioxide levels (Kuo *et al.* 1997, Sainju *et al.* 2002, Lal 2003).

Although cover crops can perform multiple functions in an agroecosystem simultaneously, they are often grown for the sole purpose of preventing soil erosion. Soil erosion is a process that can irreparably reduce the productive capacity of an agroecosystem. Dense cover crop stands physically slow down the velocity of rainfall before it contacts the soil surface, preventing soil splashing and erosive surface runoff (Romkens *et al.* 1990). Additionally, vast cover crop root networks help anchor the soil in place and increase soil porosity, creating suitable habitat networks for soil macrofauna (Tomlin *et al.* 1995).

Soil quality is managed to produce optimum circumstances for crops to flourish. The principal factors of soil quality are soil salination, pH, microorganism balance and the prevention of soil contamination.

Water Management

By reducing soil erosion, cover crops often also reduce both the rate and quantity of water that drains off the field, which would normally pose environmental risks to waterways and ecosystems downstream (Dabney *et al.* 2001). Cover crop biomass acts as a physical barrier between rainfall and the soil surface, allowing raindrops to steadily trickle down through the soil profile. Also, as stated above, cover crop root growth results in the formation of soil pores, which in addition to enhancing soil macrofauna habitat provides pathways for water to filter through the soil profile rather than draining off the field as surface flow. With increased water infiltration, the potential for soil water storage and the recharging of aquifers can be improved (Joyce *et al.* 2002).

Just before cover crops are killed (by such practices including mowing, tilling, discing, rolling, or herbicide application) they contain a large amount of moisture. When the cover crop is incorporated into the soil, or left on the soil surface, it often increases soil moisture. In agroecosystems where water for crop production is in short supply, cover crops can be used as a mulch to conserve water by shading and cooling the soil surface. This reduces evaporation of soil moisture. In other situations farmers try to dry the soil out as quickly as possible going into the planting season. Here prolonged soil moisture conservation can be problematic. While cover crops can help to conserve water, in temperate regions (particularly in years with below average precipitation) they can draw down soil water supply in the spring, particularly if climatic growing conditions are good. In these cases, just before crop planting, farmers often face a tradeoff between the benefits of increased cover crop growth and the drawbacks of reduced soil moisture for cash crop production that season.

Weed Management

Thick cover crop stands often compete well with weeds during the cover crop growth period, and can prevent most germinated weed seeds from completing their life cycle and reproducing. If the cover crop is left on the soil surface rather than incorporated into the soil as a green manure after its growth is terminated, it can form a nearly impenetrable mat. This drastically reduces light transmittance to weed seeds, which in many cases reduces weed seed germination rates (Teasdale 1993). Furthermore, even when weed seeds germinate, they often run out of stored energy for growth before building the necessary structural capacity to break through the cover crop mulch layer. This is often termed the *cover crop smother effect* (Kobayashi *et al.* 2003).

Some cover crops suppress weeds both during growth and after death (Blackshaw *et al.* 2001). During growth these cover crops compete vigorously with weeds for available space, light, and nutrients, and after death they smother the next flush of weeds by forming a mulch layer on the soil surface. For example, Blackshaw *et al.* (2001) found that when using *Melilotus officinalis* (yellow sweetclover) as a cover crop in an improved fallow system (where a fallow period is intentionally improved by any number of different management practices, including the planting of cover crops), weed biomass only constituted between 1-12% of total standing biomass at the end of the cover crop growing season. Furthermore, after cover crop termination, the yellow sweetclover residues suppressed weeds to levels 75-97% lower than in fallow (no yellow sweetclover) systems .

In addition to competition-based or physical weed suppression, certain cover crops are known to suppress weeds through allelopathy (Creamer *et al.* 1996, Singh *et al.* 2003). This occurs when certain biochemical cover crop compounds are degraded that happen to be toxic to, or inhibit seed germination of, other plant species. Some well known examples of allelopathic cover crops are *Secale cereale* (rye), *Vicia villosa* (hairy vetch), *Trifolium pratense* (red clover), *Sorghum bicolour* (sorghum-sudangrass), and species in the Brassicaceae family, particularly mustards (Haramoto and Gallandt 2004). In one study, rye cover crop residues were found to have provided between 80% and 95% control of early season broadleaf weeds when used as a mulch during the production of different cash crops such as soybean, tobacco, corn, and sunflower (Nagabhushana *et al.* 2001).

In a recent study released by the Agricultural Research Service (ARS) scientists examined how rye seeding rates and planting patterns affected cover crop production. [1] The results show that planting more pounds per acre of rye increased the cover crop's production as well as decreased the amount of weeds. The same was true when scientists tested seeding rates on legumes and oats; a higher density of seeds planted per acre decreased the amount of weeds and increased the yield of legume and oat production. The planting patterns, which consisted of either traditional rows or grid patterns, did not seem to make a significant impact on the cover crop's production or on the weed production in either cover crop. The ARS scientists concluded that increased seeding rates could be an effective method of weed control.

Disease Management

In the same way that allelopathic properties of cover crops can suppress weeds, they can also break disease cycles and reduce populations of bacterial and fungal diseases (Everts 2002), and parasitic nematodes (Potter *et al.* 1998, Vargas-Ayala *et al.* 2000). Species in the Brassicaceae family, such as mustards, have been widely shown to suppress fungal disease populations through the release of naturally occurring toxic chemicals during the degradation of glucosinolade compounds in their plant cell tissues (Lazzeri and Manici 2001).

Pest Management

Some cover crops are used as so-called "trap crops", to attract pests away from the crop of value and toward what the pest sees as a more favourable habitat (Shelton and Badenes-Perez 2006). Trap crop areas can be established within crops, within farms, or within landscapes. In many cases the trap crop is grown during the same season as the food crop being produced. The limited area occupied by these trap crops can be treated with a pesticide once pests are drawn to the trap in large enough numbers to reduce the pest populations. In some organic systems, farmers drive over the trap crop with a large vacuum-based implement to physically pull the pests off the plants and out of the field (Kuepper and Thomas 2002). This system has been recommended for use to help control the lygus bugs in organic strawberry production (Zalom *et al.* 2001). Other cover crops are used to attract natural predators of pests by providing elements of their habitat.

This is a form of biological control known as habitat augmentation, but achieved with the use of cover crops (Bugg and Waddington 1994).

Findings on the relationship between cover crop presence and predator/ pest population dynamics have been mixed, pointing toward the need for detailed information on specific cover crop types and management practices to best complement a given integrated pest management strategy. For example, the predator mite *Euseius tularensis* (Congdon) is known to help control the pest citrus thrips in Central California citrus orchards. Researchers found that the planting of several different leguminous cover crops (such as bell bean, woollypod vetch, New Zealand white clover, and Austrian winter pea) provided sufficient pollen as a feeding source to cause a seasonal increase in *E. tularensis* populations, which with good timing could potentially introduce enough predatory pressure to reduce pest populations of citrus thrips (Grafton-Cardwell *et al.* 1999).

Diversity and Wildlife

Although cover crops are normally used to serve one of the above discussed purposes, they often simultaneously improve farm habitat for wildlife. The use of cover crops adds at least one more dimension of plant diversity to a cash crop rotation. Since the cover crop is typically not a crop of value, its management is usually less intensive, providing a window of "soft" human influence on the farm. This relatively "hands-off" management, combined with the increased on-farm heterogeneity created by the establishment of cover crops, increases the likelihood that a more complex trophic structure will develop to support a higher level of wildlife diversity.

In one study, researchers compared arthropod and songbird species composition and field use between conventionally and cover cropped cotton fields in the Southern United States. The cover cropped cotton fields were planted to clover, which was left to grow in between cotton rows throughout the early cotton growing season (stripcover cropping). During the migration and breeding season, they found that songbird densities were 7–20 times higher in the cotton fields with integrated clover cover crop than in the conventional cotton fields. Arthropod abundance and biomass was also higher in the clover cover cropped fields throughout much of the songbird breeding season, which was attributed to an increased supply of flower nectar from the clover. The clover cover crop enhanced songbird habitat by providing cover and nesting sites, and an increased food source from higher arthropod populations

5

Disease Causes and Controls

Diseases of vegetables grown in the home garden may reduce both the yield and quality of vegetables. Controlling such diseases often determines success or failure and adds immeasurably to the pleasure derived from the garden. Diseases of vegetables are caused by microorganisms, including fungi, bacteria (including mycoplasmas and spiroplasmas), viruses, and nematodes. These organisms are spread by wind-blown or water-splashed spores, insects, infected seeds or transplants, by the movement of infested soil, and by humans handling wet plants.

Vegetable diseases can be successfully controlled using an integrated disease control program. Such a program involves the use of crop rotation, sanitation, disease-resistant or tolerant varieties, disease-free seeds or transplants, sound horticultural practices, balanced soil fertility (based on soil tests), and proper and timely applications of pesticides. There is no single control practice that can control all of the diseases of any vegetable crop. Therefore, several measures must be used to achieve satisfactory control.

Disease Control before the Garden is Planted

Choose and prepare the site. The site for the vegetable garden should be well-drained, have adequate sunlight (8 or 10 hours of direct sunshine per day), and good air circulation. The soil should have adequate soil fertility and the proper pH (5.5 to 7.0). Information on garden location, soil fertility, and soil pH can be found in University of Illinois Extension Circular 1331, Vegetable Gardening in the

Midwest. If the area has produced garden vegetables for several years, sanitation and rotation practices will reduce the risk of disease-causing organisms that have survived from preceding crops. Removing or turning under dead and diseased plant material in the fall is important for disease control. Crop residues from healthy plants can be composted and returned to the garden. Disease-causing fungi and bacteria can survive composting unless the decomposition of the crop residue is complete, a situation that seldom occurs. Therefore, we suggest that no diseased plant parts be placed in the compost pile. Some disease organisms will even survive complete composting, including corn smut; clubroot of cabbage, broccoli, cauliflower and other crucifers; Verticillium wilt of potato, tomato, pepper, and eggplant; and root-knot nematodes. Turning (plowing or tilling) under crop residues in the fall helps prevent the over-wintering of many disease-causing organisms. Besides destroying the “winter home” of various organisms, this practice also helps control the insects that transmit certain diseases. Also, crop debris that is turned under improves soil tilth and makes spring garden work easier.

Crop rotation. Crop rotation is an important control measure since many disease-causing organisms only attack related plants in the same family. Unrelated plants do not serve as hosts on which these organisms can multiply. Therefore, if possible, avoid planting any vegetables within each of the following groups in the same location more than once every three or four years.

Cole crops–(cabbage family) broccoli, Brussels sprout, cabbage, cauliflower, Chinese cabbage, mustard greens, kohlrabi, radish, rutabaga, and turnip. Cucurbits–(cucumber family) cucumber, gourds, muskmelon, pumpkin, squash (all types), and watermelon.

Solanaceous crops–(tomato family) eggplant, husk tomato (ground cherry), potato, pepper, tomato. Chenopodiaceous crops–beet (red and sugar) and spinach.

Amaryllidaceae–(onion family) chives, leeks, onions, shallots, and garlic.

Legumes–pea, beans (all types). Umbelliferous crops–carrot, parsnip, celery. For example, cabbage, broccoli, radish, or turnips should not be planted in the same location for two successive years. Cabbage could be followed with beans, a vine crop such as cucumber, or with sweet corn.

Diseases affecting beans, peas, sweet corn, carrot, lettuce, spinach, rhubarb, okra, edible soybeans, and similar plants are usually specific to only one of these crops, and generally will not infect others or members of the major vegetable crop groups.

Choose disease-resistant varieties. The use of well-adapted, disease-resistant varieties is the simplest and most efficient method of controlling many diseases. Lists of disease-resistant vegetables recommended for use by Illinois gardeners are given in Circular 1331 (Vegetable Gardening in the Midwest. Additional information on disease resistance is also given in the following paragraphs. A listing of varieties and their resistance can be found in Circular 1363, Midwest Vegetable Production Guide for Commercial Growers.

Start with disease-free seeds, plants, and planting materials. It is important to start with disease-free seeds or planting materials (bulbs, tubers, transplants, sets, and so on) to keep from introducing serious diseases into the garden. In general, gardeners are encouraged not to save their own seed but to purchase seed from reputable seed dealers. Hot-water soaks and fungicide seed treatments may be used to control the seedborne disease organisms that cause seed rots and damping-off. Since temperature controls and timing must be precise, home gardeners usually buy seed that has already been treated with hot water by the seed processor. Hot-water treatment of cabbage, broccoli, Brussels sprout, and cauliflower seed is particularly important.

Many vegetable seeds are coated with a fungicide at the time of purchase, as evidenced by the coloured coating on the seed. Untreated seed can be treated by the home gardener by placing a small quantity (the size of one or two match heads) of captan or thiram in the seed packet and shaking the packet for a minute or two to thoroughly cover the seed. Excess protectant may be sifted out before planting.

Do not plant diseased material (for example, transplants, sets, bulbs, or tubers). All planting material should be healthy and free of yellowing and brown or black spots, and should not be stunted or show poor development. Only certified, disease-free potato tubers or sweet potato slips should be used. Examine transplants thoroughly for signs of leaf or stem disease. It never pays to buy and plant diseased transplants, no matter what the price.

Damping-off diseases affecting home-grown vegetable transplants can be controlled by the use of diseasefree soil and fungicides.

Cultural Practices

Cultural practices such as controlling weeds, planting at the right time, planting at the proper depth and spacing, employing cages or stakes, and watering when the temperature is rising to promote a rapid drying of the foliage can help control many foliar diseases. Mulches help to control fruit rots and aid in control of blossom-end rot of tomato, pepper, and vine crops by maintaining a uniform supply of moisture in the soil. Weed control in and around the garden reduces the risk of attacks by viruses that overwinter in these plants. Do not work or harvest wet plants, since this can spread many diseases, particularly those caused by bacterial and fungal pathogens.

Disease Control During the Growing Season

Many diseases, such as the early blight disease of potato and tomato, occur each year despite all preplanting precautions. For such diseases, applications of fungicides and bactericides to the growing plants may be needed.

The best way to apply these materials is as sprays before the disease occurs. Often a wetting agent such as liquid detergent or soap (1/2 teaspoon in 1 gallon) is added to obtain more thorough wetting and coverage of the foliage. Apply sprays to the point of run-off on a 7- to 10-day schedule. This maintains a fresh or effective covering of fungicide and protects the new growth. Fungicides and bactericides currently recommended for use on vegetables is given in the Illinois Pest Control Handbook which is updated each year.

Principal Diseases of Vegetables and Control Measures

Asparagus: Rust. Grow resistant varieties Start spray applications after harvest and continue until mid-August on a 10-day schedule with an approved fungicide.

Crown and root rots. Maintain proper fertility and good soil drainage. Avoid over-cutting.

Beans: Seed decay and damping-off. Use captan- or thiram-treated seed. Plant in a warm, moist, well-prepared seedbed.

Bacterial Blights: Do not save seed from infected plants. Purchase seed from a reputable seed dealer. Spray a copper-based bactericide such as copper sulfate at the first sign of disease, and continue on a 7- day scheduled until harvest. Do not work among wet plants.

Rust: Grow rust-resistant varieties. Apply fungicides at the first sign of disease and continue on a 7-day schedule.

Mosaics: Grow resistant varieties..

Beets: Seed decay and damping-off. Plant seed treated with thiram or captan.

Leaf Spots: Spray with an approved fungicide on a 7-day schedule.

Carrots

Seed Rot and Damping-Off: Plant seed treated with captan or thiram.

Leaf Spots: Practice a 2- to 3-year crop rotation. Spray weekly with an approved fungicide, starting about mid-June.

Aster Yellows: Destroy infected plants as soon as they appear. Apply insecticide sprays for leafhopper control. Spray before removing infected plants.

Cole Crops: (cabbage, broccoli, Brussels sprout, cauliflower, Chinese cabbage, kohlrabi, mustard greens, radish, rutabaga, and turnip).

Seed Decay, Damping-Off, Black Rot, and Blackleg: Plant hot-water-treated seed which has also been treated with captan or thiram. Several black-rot-resistant cabbage varieties are now available.

Leaf Spots and Blights: Practice 2- to 3-year crop rotation for broccoli, cabbage, cauliflower, and Brussels sprout. Apply an approved fungicide on a 7-day schedule.

Clubroot: Purchase disease-free transplants. PCNB (Terraclor) may be added to the transplant water.

Fusarium Yellows: Grow only varieties that are resistant to yellows. Cucurbits (Cucumber, gourds, muskmelon, pumpkin, squash, and watermelon).

Seed Rot and Damping-Off: Plant seed treated with captan. Plant in a warm, well-prepared seed bed.

Angular leafspot: Practice a 2- to 3-year crop rotation. Spray with a fixed copper bactericide, such as COCS. Remove diseased plant material. Plant resistant varieties whenever possible.

Anthracnose, Alternaria Leaf Blight, Downy Mildew, Blossom Blights, Fruit Rots and Spots, and Gummy Stem Blight or Black Rot: Practice 2- to 3-year crop rotation. Apply an approved fungicide at weekly intervals, starting when the plants are in the two-leaf stage. Adequate nitrogen fertility is essential for the successful control of Alternaria leaf blight. Plant resistant varieties when available.

Bacterial Wilt: Spray with an insecticide to control cucumber beetles as soon as the seedlings "crack" the soil before they emerge. Continue weekly spraying until the plants are in bloom. Spray in the evening to avoid killing bees.

Fusarium Wilt: Grow only cantaloupe and watermelon varieties that are resistant to Fusarium wilt.

Powdery Mildew: Spray with an approved fungicide at the first sign of disease, and weekly thereafter. Grow resistant varieties.

Mosaic and other virus diseases. Grow resistant varieties where possible. Only cucumber-mosaicresistant cucumbers should be grown.

Eggplant: Seed rot, damping-off, and Phomopsis blight. Plant seed treated with captan or thiram.

Phomopsis Blight and Other Fruit Rots: Practice a 2- to 3-year crop rotation. Spray weekly with an approved fungicide when fruits are half size, or when disease first appears.

Verticillium Wilt: Mulching with black plastic may help reduce losses.

Onion: Smut, seed rot, and damping-off. Plant seed treated with thiram or captan.

Leaf Diseases: Practice a 2- to 3-year crop rotation. Spray weekly with an approved fungicide. Carefully follow label directions as regards the days between the last spray and harvest.

Bulb Rots: Control leaf diseases. Also follow these recommendations: avoid storing improperly cured or injured bulbs; let all onion tops fall over (ripen) naturally; store under cold dry conditions; and do not allow cured bulbs to be exposed to sunscald, water, or high humidity.

Fusarium Basal Rot: Grow resistant varieties.

Pea: Seed rot and damping-off. Plant seed treated with captan or thiram.

Root Rot: Plant early in well-drained soils. Use a fungicide seed treatment.

Fusarium Wilt: Plant only varieties that are resistant to wilt.

Peppers: Seed rot and damping-off. Plant seed treated with captan or thiram.

Bacterial Spot: Use the hot-water seed treatment, or dip seed in a solution of 1 part household bleach to 3 parts water and rinse with water. Follow with a fungicide seed treatment. Apply weekly sprays of a fixed copper fungicide such as COCS, through the bloom stage. Use crop rotation outside the tomato family

Tobacco Mosaic: Plant resistant varieties.

Potato: Viruses, late blight, Verticillium wilt, and ring rot. Plant only "certified" disease-free seed.

Seedpiece Rots: Treat cut seedpieces with captan as a dust or dip. Plant in warm soil (over 50°F) after the cut surfaces have corked over (suberized).

Early Blight and Late Blight: Practice a 2- to 3-year crop rotation. Destory or remove cull tubers. Spray at weekly intervals, starting when the plants are 4 to 6 inches high with an approved fungicide. Plant late blight-resistant varieties.

Verticillium Wilt and Root-Knot Nematodes: Plant resistant varieties.

Scab: Plant resistant varieties, or adjust the soil pH to 5.2 to 5.5. Use fungicide seedpiece treatments.

Storage Rots: Store only disease-free, unbruised tubers. Allow tubers to cure at 60°F for 5 to 7 days before reducing temperature to 36° to 40°F (if possible).

Sweet Corn

Seed Rot and Damping-Off: Plant seed treated with captan or thiram.

Smut: Plant resistant varieties. Remove smut galls from the garden before they break. Do not compost smutted corn or infected corn residues.

Stewart's Bacterial Wilt: Plant resistant varieties. Control flea beetles with a recommended insecticide. Early season control is particularly important.

Foliage Blights and Rust: Plant varieties resistant to northern and southern corn leaf blights and rusts. Apply an approved fungicide. However, the use of fungicides is rarely justified. Diseased plant material should be removed from the garden. Blighted leaf tissue is safe to compost.

Maize Dwarf Mosaic: Plant as far from Johnsongrass as possible. Control wild Johnsongrass plants. Plant tolerant varieties.

Spinach

Seed Rot and Damping-Off: Plant seed treated with captan or thiram.

Cucumber Mosaic Virus (Blight): Plant resistant varieties.

Downy Mildew or Blue Mold: Plant resistant varieties.

Sweet Potato

Black Rot, Scurf, Foot Rot, Wilt, and Soil Rot: Buy certified, disease-free slips. Plant resistant varieties. Use a 3- to 4-year rotation.

Storage Diseases: Do not expose roots to temperatures below 55°F. Cure the roots immediately after digging at 85°F and high humidity for 10 to 14 days. Store at 55°F. Never store in airtight bags or containers. Apply an approved fungicide as a post-harvest dip. Follow the manufacturer's directions carefully.

Tomato

Seed Rot, Damping-off, Bacterial Spot, Bacterial Canker, and Tobacco Mosaic Virus: Plant seed which has been treated with hot water or acid and then treated with captan or thiram. Purchase seed or transplants only from reputable dealers. Soaking seed in copper sulfate solution (2 ounces per gallon) or dipping in bleach solution as described under peppers will help prevent bacterial spot and canker if seed are saved.

Bacterial Spot. Plant Certified, Disease-Free Seed or Transplants: Practice a 2-to 3-year crop rotation. Spray plants weekly with a fixed copper fungicide, beginning with the first sign of disease, through the bloom stage.

Early Blight, Septoria Leaf Blight, Anthracnose, Buckeye Rot, and Soil Rots: Spray weekly with an approved fungicide. Practice 2- to 3-year crop rotation. Remove or destroy infected plant parts. Carefully follow label directions as regards the days between the last spray and harvest. Staking or caging plants and providing good air circulation will help reduce foliar diseases. Staking, caging or mulching will also help prevent fruit rots, such as anthracnose, buckeye rot, and soil rot.

Blossom-End Rot: Maintain uniform soil moisture by mulching the plants and irrigating when needed. Avoid heavy pruning, excessive nitrogen applications, or damage to the roots by cultivation or hoeing.

Verticillium and Fusarium Wilts: Grow only VF - or VFN-resistant varieties. There are many varieties available.

Root-Knot Nematodes: Grow VFN-resistant varieties.

Viruses: Avoid contact between potatoes and tomatoes to prevent "double infections" of potato and tomato viruses. Wash hands thoroughly with soap and hot water before handling your plants. Do not use tobacco when working with tomatoes.

Asparagus and Artichoke

Asparagus officinalis is a spring vegetable, a flowering perennial plant species in the genus *Asparagus*. It was once classified in the lily family, like its *Allium* cousins, onions and garlic, but the Liliaceae have been split and the onion-like plants are now in the family Amaryllidaceae and asparagus in the Asparagaceae. *Asparagus officinalis* is native to most of Europe, northern Africa and western Asia, and is widely cultivated as a vegetable crop.

Biology

Asparagus is a herbaceous, perennial plant growing to 100–150 centimetres (39–59 in) tall, with stout stems with much-branched feathery foliage. The "leaves" are in fact needle-like cladodes (modified stems) in the axils of scale leaves; they are 6–32 mm (0.24–1.3 in) long and 1 mm (0.039 in) broad, and clustered 4–15 together. The root system is adventitious and the root type is fasciculated. The flowers are bell-shaped, greenish-white to yellowish, 4.5–6.5 mm (0.18–0.26 in) long, with six tepals partially fused together at the base; they are produced singly or in clusters of two or three in the junctions of the branchlets. It is usually dioecious, with male and female flowers on separate plants, but sometimes hermaphrodite flowers are found. The fruit is a small red berry 6–10 mm diameter, which is poisonous to humans.

Plants native to the western coasts of Europe (from northern Spain north to Ireland, Great Britain, and northwest Germany) are treated as *Asparagus officinalis* subsp. *prostratus* (Dumort.) Corb., distinguished by its low-growing, often prostrate stems growing to only 30–70 cm (12–28 in) high, and shorter cladodes 2–18 mm (0.079–0.71 in) long. It is treated as a distinct species, *Asparagus prostratus* Dumort, by some authors. A remarkable adaptation is the edible asparagus, while in the Macaronesian Islands several species, (*A. umbellatus*, *A. scoparius*, etc.), are preserved the original form, a leafy vine; in the Mediterranean, the asparagus genus has evolved into thorny species.

History

Asparagus has been used as a vegetable and medicine, owing to its delicate flavour, diuretic properties, and more. It is pictured as an offering on an Egyptian frieze dating to 3000 BC. Still in ancient times, it was known in Syria and in Spain. Greeks and Romans ate it fresh when in season and dried the vegetable for use in winter;

Romans would even freeze it high in the Alps, for the Feast of Epicurus. Emperor Augustus reserved the "Asparagus Fleet" for hauling the vegetable, and coined the expression "faster than cooking asparagus" for quick action. A recipe for cooking asparagus is in the oldest surviving book of recipes, Apicius's third-century AD *De re coquinaria*, Book III.

The ancient Greek physician Galen (prominent among the Romans) mentioned asparagus as a beneficial herb during the second century AD, but after the Roman empire ended, asparagus drew little medieval attention, until al-Nafzawi's *The Perfumed Garden.* That piece of writing celebrates its (scientifically unconfirmed) aphrodisiacal power, a supposed virtue that the Indian *Ananga Ranga* attributes to "special phosphorus elements" that also counteract fatigue. By 1469, asparagus was cultivated in French monasteries. Asparagus appears to have been hardly noticed in England until 1538, and in Germany until 1542.

The finest texture and the strongest and yet most delicate taste is in the tips. The *points d'amour* ("love tips") were served as a delicacy to Madame de Pompadour. Asparagus became available to the New World around 1850, in the United States.

Uses

Only young asparagus shoots are commonly eaten: once the buds start to open ("ferning out"), the shoots quickly turn woody.

Water makes up 93% of Asparagus's composition. Asparagus is low in calories and is very low in sodium. It is a good source of vitamin B_6, calcium, magnesium and zinc, and a very good source of dietary fibre, protein, beta-carotene, vitamin C, vitamin E, vitamin K, thiamin, riboflavin, rutin, niacin, folic acid, iron, phosphorus, potassium, copper, manganese and selenium, as well as chromium, a trace mineral that enhances the ability of insulin to transport glucose from the bloodstream into cells. The amino acid asparagine gets its name from asparagus, as the asparagus plant is relatively rich in this compound.

The shoots are prepared and served in a number of ways around the world, typically as an appetizer or vegetable side dish. In Asian-style cooking, asparagus is often stir-fried. Cantonese restaurants in the United States often serve asparagus stir-fried with chicken, shrimp, or beef. Asparagus may also be quickly grilled over charcoal or hardwood embers. It is also used as an ingredient in some stews and soups. In recent years, almost as a cycle dating back to early culinary habits, asparagus has regained its popularity eaten raw as a component of a salad.

Asparagus can also be pickled and stored for several years. Some brands may label shoots prepared this way as "marinated".

Stem thickness indicates the age of the plant, with the thicker stems coming from older plants. Older, thicker stalks can be woody and peeling the skin at the base will remove the tough layer. Peeled asparagus will poach much faster, however. The bottom portion of asparagus often contains sand and dirt, so thorough cleaning is generally advised before cooking.

Green asparagus is eaten worldwide, though the availability of imports throughout the year has made it less of a delicacy than it once was. In Europe, however, the "asparagus season is a highlight of the foodie calendar"; in the UK this traditionally begins on 23 April and ends on Midsummer Day. As in continental Europe, due to the short growing season and demand for local produce, asparagus commands a premium.

White Asparagus in Continental Northwestern Europe

Asparagus is very popular in the Netherlands, Spain, France, Poland, Belgium, Germany, Italy and Switzerland, and is almost exclusively white—else, it is specified by the local language term for "green asparagus". Compared to the latter, the locally cultivated so-called "white gold" or "edible ivory", also referred to as "the royal vegetable", is less bitter and much more tender. Freshness is highly regarded, and it must be peeled before cooking or raw consumption.

To cultivate white asparagus, the shoots are covered with soil as they grow; without exposure to sunlight there is no photosynthesis, and the shoots remain white in colour.

Only seasonally on the menu, asparagus dishes are advertised outside many restaurants, usually from late April to June. For the French style, asparagus is often boiled or steamed and served with hollandaise sauce, melted butter or olive oil, Parmesan cheese or mayonnaise. Tall, narrow asparagus cooking pots allow the shoots to be steamed gently, their tips staying out of the water.

During the German *Spargelsaison* or *Spargelzeit*, the asparagus season that traditionally finishes on 24 June, roadside stands and open-air markets sell about half of the country's white asparagus consumption.

Medicinal

The second century physician Galen described asparagus as "cleansing and healing".

Nutrition studies have shown asparagus is a low-calorie source of folate and potassium. Its stalks are high in antioxidants. "Asparagus provides essential nutrients: six spears contain some 135 micrograms (ìg) of folate, almost half the adult RDI (recommended daily intake), 20 milligrams of potassium," notes an article in *Reader's Digest.*

Research suggests folate is key in taming homocysteine, a substance implicated in heart disease. Folate is also critical for pregnant women, since it protects against neural tube defects in babies. Studies have shown that people who have died from Alzheimer's Disease have extremely low to no levels of folate. Several studies indicate getting plenty of potassium may reduce the loss of calcium from the body.

Particularly green asparagus is a good source of vitamin C. Vitamin C helps the body produce and maintain collagen, the major structural protein component of the body's connective tissues.

"Asparagus has long been recognised for its medicinal properties," wrote D. Onstad, author of *Whole Foods Companion: A Guide for Adventurous Cooks, Curious Shoppers and Lovers of Natural Foods.* "Asparagus contains substances that act as a diuretic, neutralize ammonia that makes us tired, and protect small blood vessels from rupturing. Its fiber content makes it a laxative, too."

Water from cooking asparagus may help clean blemishes on the face if used for washing the face morning and night. From John Heinerman's *Heinerman's New Encyclopedia of Fruits and Vegetables*: "Cooked asparagus and its watery juices are very good for helping dissolve uric acid (causes gout) deposits in the extremities, as well as inducing urination where such a function may be lacking or only done on an infrequent basis.

Asparagus is especially useful in cases of hypertension where the amount of sodium in the blood far exceeds the potassium present. Cooked asparagus also increases bowel evacuations."

South Korean scientists discovered asparagus can help with hangovers. Research to be published in the Journal of Food Science says extracts taken from leaves and shoots were found to boost levels of key enzymes that help break down alcohol.

Cultivation

Since asparagus often originates in maritime habitats, it thrives in soils that are too saline for normal weeds to grow. Thus, a little salt was traditionally used to suppress weeds in beds intended for asparagus; this has the disadvantage that the soil cannot be used for anything else. Some places are better for growing asparagus than others. The fertility of the soil is a large factor. "Crowns" are planted in winter, and the first shoots appear in spring; the first pickings or "thinnings" are known as sprue asparagus. Sprue has thin stems.

Figure: *Green asparagus for sale in New York City*

A new breed of "Early Season Asparagus" that can be harvested two months earlier than usual was announced by a UK grower in early 2011. This variety does not need to lie dormant and blooms at 7 °C (45 °F) rather than the usual 9 °C (48 °F).

The blanching of white asparagus is obtained by the labour intensive hilling cultivation method, to distinguish its gastronomical qualities from those of the green plant, which is the same botanical variety.

Purple asparagus differs from its green and white counterparts, having high sugar and low fiber levels. Purple asparagus was originally developed in Italy and commercialized under the variety name "Violetto d' Albenga". Since then, breeding work has continued in countries such as the United States and New Zealand.

Companion Planting

Asparagus is a useful companion plant for tomatoes. The tomato plant repels the asparagus beetle, as do several other common companion plants of tomatoes. Meanwhile, asparagus may repel some harmful root nematodes that affect tomato plants.

Celebrations

The green crop is significant enough in California's Sacramento-San Joaquin River Delta region that the city of Stockton holds a festival every year to celebrate it, as does the city of Hart, Michigan, complete with a parade and asparagus queen. The Vale of Evesham in Worcestershire is heralded as the largest producer within Northern Europe, celebrating like Stockton, with a week-long festival every year involving auctions of the best crop and locals dressing up as spears of asparagus as part of the British Asparagus Festival.

Many German cities hold an annual *Spargelfest* (asparagus festival) celebrating the harvest of white asparagus. Schwetzingen claims to be the "Asparagus Capital of the World" and during its festival an Asparagus Queen is crowned. The Bavarian city of Nuremberg feasts a week long in April, with a competition to find the fastest asparagus peeler in the region. This usually involves generous amounts of the local wines and beer being consumed to aid the spectators' appreciative support.

Vernacular Names and Etymology

Figure: *Mature native asparagus with seed pods in Saskatchewan, Canada*

Asparagus officinalis is widely known simply as "asparagus", and may be confused with unrelated plant species also known as "asparagus", such as *Ornithogalum pyrenaicum* known as "Prussian asparagus" for its edible shoots.

The English word "asparagus" derives from classical Latin, but the plant was once known in English as *sperage*, from the Medieval Latin *sparagus*. This term itself derives from the Greek *aspharagos* or *asparagos*, and the Greek term originates from the Persian *asparag*, meaning "sprout" or "shoot". Asparagus was also corrupted in some places to "sparrow grass"; indeed, the Oxford English Dictionary quotes John Walker as having written in 1791 that "*Sparrow-grass* is so general that *asparagus* has an air of stiffness and pedantry". In Gloucestershire and Worcestershire, it is also known simply as "grass", and young plants too small to cut are called "pru". Another known colloquial variation of the term, most common in parts of Texas, is "aspar grass" or "asper grass". In the Midwest United States and Appalachia, "spar grass" is a common colloquialism. Asparagus is commonly known in fruit retail circles as "Sparrows Guts", etymologically distinct from the old term "sparrow grass", thus showing convergent language evolution.

The Sanskrit name of Asparagus is *shatavari* and it has been historically used in India as a part of Ayurvedic medicines. In Kannada, it is known as *ashadhi*, *majjigegadde* or *sipariberuballi*. In China it is known as lu sun (*Cantonese*: lo sun) †M{ (simplified ¦,{), in Thailand as *no mai farang* (Thai: Ë¹èÍäÁé½ÃÑè§), and in Vietnam as *măng tây* which literally mean "European bamboo shoots" and "Western bamboo shoots", respectively. The green asparagus is commonly used in Chinese-American cuisine and Thai cuisine.

Effects on Urine

The effect of eating asparagus on the urine has long been observed:

> *"[Asparagus] cause a filthy and disagreeable smell in the urine, as every Body knows."* (Treatise of All Sorts of Foods, *Louis Lemery, 1702) "asparagus... affects the urine with a foetid smell (especially if cut when they are white) and therefore have been suspected by some physicians as not friendly to the kidneys; when they are older, and begin to ramify, they lose this quality; but then they are not so agreeable." ("An Essay Concerning the Nature of Aliments," John Arbuthnot, 1735)*

"A few Stems of Asparagus eaten, shall give our Urine a disagreable Odour..." ("Letter to the Royal Academy of Brussels," Benjamin Franklin, c. 1781)

Asparagus "...transforms my chamber-pot into a flask of perfume." Marcel Proust (1871–1922)

There is debate about whether all—or only some—people produce the smell, and whether all (or only some) people identify the smell. It was originally thought this was because some of the population digested asparagus differently from others, so some people excreted odorous urine after eating asparagus, and others did not. In the 1980s three studies from France, China and Israel published results showing that producing odorous urine from asparagus was a common human characteristic. The Israeli study found that from their 307 subjects all of those who could smell 'asparagus urine' could detect it in the urine of anyone who had eaten asparagus, even if the person who produced it could not detect it himself. However, a 2010 study found variations in both production of odorous urine and the ability to detect the odour, but that these were not tightly related. It is believed most people produce the odorous compounds after eating asparagus, but only about 22% of the population have the autosomal genes required to smell them.

In 2010, the company 23andMe published a genome-wide association study on whether participants have "ever noticed a peculiar odor when you pee after eating asparagus?" This study pinpointed a single-nucleotide polymorphism (SNP) in a cluster of olfactory genes associated with the ability to detect the odor. While this SNP did not explain all of the difference in detection between people, it provides support for the theory that there are genetic differences in olfactory receptors that lead people to be unable to smell these odorous compounds.

Chemistry

Certain compounds in asparagus are metabolized to yield ammonia and various sulfur-containing degradation products, including various thiols and thioesters, which give urine a characteristic smell.

Some of the volatile organic compounds responsible for the smell are:

- methanethiol
- dimethyl sulfide
- dimethyl disulfide
- bis(methylthio)methane
- dimethyl sulfoxide
- dimethyl sulfone

Subjectively, the first two are the most pungent, while the last two (sulfur-oxidized) give a sweet aroma. A mixture of these compounds form a "reconstituted asparagus urine" odor. This was first investigated in 1891 by Marceli Nencki, who attributed the smell to methanethiol. These compounds originate in the asparagus as asparagusic acid and its derivatives, as these are the only sulfur-containing compounds unique to asparagus. As these are more present in young asparagus, this accords with the observation that the smell is more pronounced after eating young asparagus. The biological mechanism for the production of these compounds is less clear.

The onset of the asparagus urine smell is remarkably rapid. The smell has been reported to be detectable 15 to 30 minutes after ingestion.

Crop Maintenance

Asparagus ferns (also called brush) produce food which is stored in the roots and determines the yields of spears the next year. Weed, insect, and disease control and proper fertilization are all important for good fern growth and good yields. Cultivate very shallowly since serious injury to the roots may result from deep cultivation. Ferns will not be killed by the first frost and should be allowed to grow until it turns brown (usually in early December).

Fertilization - During succeeding years apply 50 to 75 pounds per acre of nitrogen and 75 to 125 pounds of potassium twice each year. Make one application just before first appearance of spears in the spring in early March. Apply the same amount of fertilizer at the conclusion of the harvest season in mid-May. Thus 100 to 150 pounds of nitrogen and 140 to 250 pounds of potassium will be applied each year. Apply the fertilizer to the top of the soil or with very shallow incorporation. Growers may wish to use stable manure to increase soil humus. Manure can be beneficial, but is not necessary. Do not use in excess 3 to 5 tons per acre of chicken manure, or 6 to 10 tons per acre of swine or cow manure. Over-fertilization with manure can result in excessive nitrogen and poor growth.

Weed Management - Weed control in asparagus production is very important. Timely cultivation is a critical part of any asparagus weed control program, especially during the first two years. The first year asparagus should be cultivated at least once a month until September, or 6 times. The number of cultivations may be reduced by using herbicides. Weed control for asparagus is detailed in the NC Ag Chem Manual. Remove all weeds that are present after harvest. Use only very shallow disking (1 to 2 inches) to remove these weeds. Deeper disking will damage crowns and can drastically reduce yield.

Disease Management - Asparagus is affected by many diseases, of which the first 4 can be devastating if appropriate precautions are not used. If precautions are followed, the diseases pose no barrier to good crop production.

Asparagus is affected by two Fusarium diseases. The first is *Fusarium oxysporum* v. *asparagi* and is called crown rot, and *Fusarium moniliforme* which is called root rot. The root rot pathogen may be spread on asparagus seed. The crown rot fungus is found in most soils but at very low levels. If asparagus crowns are planted that have crown rot they will not produce for more than 4 to 7 years, and this is not profitable. Do not plant asparagus in soil in which asparagus has been grown in the last 5 years. Do not purchase crowns grown in soil where asparagus was grown in the last 5 years. Purchase ONLY certified crowns.

Cercospora blight (*Cercospera asparagi*) is a disease that causes the needles to fall from the mature fern. It can be identified by the browning of needles. The disease occurs when the humidity and temperature is very high. Spray with a fungicide with a HIGH PRESSURE SPRAYER (300 psi) every 7 to 10 days until the disease is controlled.

Harvested asparagus is susceptible to bacterial soft rot (*Erwinia* spp.). Wash water and water used for hydrocooling should contain chlorine. Management is based on avoiding injuries and immediately cooling harvested spears.

Rust (*Puccinia asparagi*) is not common in eastern North Carolina. The most effective way to control asparagus rust is to plant resistant varieties. Rust causes small brown rusty spots on spears and fern branches. Southern blight (*Sclerotium rolfsii*) is widespread on other crops, but is not a problem for asparagus. Asparagus is resistant to nematodes and does not require a nematicide.

Insect Management - Asparagus beetles and asparagus beetle larvae attack spears and ferns. Spray Sevin at 1 to 1.5 lb active per acre at first observation to control this insect. Armyworms can be especially bad on young ferns but can be controlled by Lannate or Nurdin at 0.9 pound active ingredient per acre. European asparagus aphid can be a problem some years. This blue-green aphid forms colonies some years in August or September. When the aphid forms colonies, it causes “Christmas tree” or bonsai effect; the new fern becomes shortened or stunted and new needles look like they are clustered. The entire fern takes on a blue-green colour.

Asparagus Diseases

Asparagus diseases, along with other garden plant ailments, are very frustrating as a gardener. It’s so disappointing to have seen beautiful fresh green buds sprouting in the spring only to find that by mid summer your roses have black spot and rust, your gooseberries have mildew or, perish the thought, your asparagus ferns are looking less than healthy!

The three main diseases affecting asparagus are:

1. Rust - cased by Puccinia asparagi
2. Fusarium stem and crown rot - caused by Fusarium moniliforme, and / or F.oxysporum f. sp. aspargi
3. Fusarium wilt and root rot - caused by Fusarium oxysporum f.sp. asparagi

Lets Just Call Them Rust or Rot!

All three diseases are fungal infections and over time will weaken your plants year on year to a greater or lesser extent depending on how well you control the spread.

An important point to remember when considering the effect of these diseases is that the strong vigorous growth of stem and fern that occurs through the summer enables carbohydrate to be channelled and stored in the roots of your crowns over the winter.

It is this store of energy that powers the growth of your plants in the spring. If your ferns are healthy and happy in the summer you will get lots of good healthy spears in the spring. By contract if your ferns are stunted by disease your spear production in the spring will be less vigorous.

Symptoms of Asparagus Diseases - What to Look Out for

Rust: Rust has three stages each looking different. Starting early in spring to early summer the disease shows itself for the first time as 10-20mm oval lesions (a bit raised) light green in colour turning creamy / orange and more sunken in the centre as they develop. They may not be that obvious and there will be more towards the base of the stems than towards the top.

The next stage is when the rust colour develops. This occurs during the summer when the 1st stage of lesions burst and spread rust coloured spores into the air to land on other plants and other parts of the same plant causing further infection. If you rub your hand over your ferns and come away with a reddish tinge to your hand you've got this problem.

Finally black, over-wintering spores are produced although these may occur at the same time as the rust coloured ones.

The rust spores which land on the feathery tips will give rust coloured spots on the fronds which will grow back down the stems into browny/ red streaks which will weaken the stems growth and may can cause the whole stem to die back.

Stem & Crown Rot

If your asparagus ferns start to look rather sad, go brown and die back before the autumn and you subsequently find that the roots are going black you probably have crown rot

Wilt & Root Rot

Again sad looking ferns that are wilting and generally in poor health will indicate this disease.

Prevention of Asparagus Diseases

If you understand a few things about these fungal diseases it will help you decide what preventative measures to take:

Even if you are starting an asparagus bed in a previously unused location away from other asparagus the fungi can be introduced from diseased crowns. You can try dipping your asparagus crowns in a fungicide solution before planting. Equally seeds can be surface contaminated with spores. An Asparagus Information Bulletin (no 202) issued by the Cooperative Extension at Cornell University suggest disinfecting seed for 2 minutes in a solution of 1 part commercial 5.25% sodium hypochloride to 4 parts water, rinsing, drying and planting immediately unless pre-soaking is planned.

The spores live over the winter on plant debris so always cut back the ferns as they die back for the winter, remove them from your vegetable patch and burn them. Take a look at our page on

Aspargus Care if you are not familiar with this process. Spores may also overwinter in the soil. Try not to plant a new bed in the same place as an old one or the spores that took a while to build up in your old bed as it matured will be waiting to spring into action on your new bed whilst it is still young.

The spores infect plants more easily when the plants are damp either as a result of dew or rain. Try to chose a spot for your plants that it not only sunny but has a breeze running through it to keep your plants from being overly damp.

If you know there are other asparagus beds around or you are planting over an old bed (because you have no choice) chose one of the more resistant varieties. Take a look at our Asparagus Plant Page for information on the disease resistance of different varieties of asparagus plant.

Fungi thrive in the damp and asparagus beds that hold too much moisture may lead to the "rots" setting in. If you have planted your bed on heavy soil which is holding a lot of moisture you could well get this problem. Generally plants that have rust are more susceptible to other asparagus diseases so treat your plants well for rust to help prevent rot!

On top of that if your plants are infested with asparagus beetle they will also be more susceptible to rust and rot.

Treating Asparagus Diseases

Even if you have taken all of these preventative measures your plants are still likely to end up suffering from one of these asparagus diseases at some stage in their life. So then what?

Early treatment is the key. Several fungicides are available for the control of rust. If your bed is in its first year or two and not ready for cutting yet you need to start spraying early in the spring, as soon as you have shoots, to prevent the spores from infecting the plants in the first place. Keep an eye out and re-spray as frequently as the products instructions suggest. If your beds are in their cutting years the act of cutting will keep the beds clear of rust unit you start leaving them to grow. As soon as you have taken your last cut is the time to start spraying, again you are aiming to stop the initial infection.

If you are getting rot to any great extent your best bet is to pray for the sort of weather that enables you to keep your beds at the right moisture level. This means not too wet. If you water in the morning this allows excess water to drain away and / or dry up during the day.

Artichoke

The globe artichoke (*Cynara carduncului var. scolymus*) is a variety of a species of thistle cultivated as a food. The edible matter is buds that form within the flower heads before the flowers come into bloom. The buds go away or change to a coarse, barely edible form when the flower blooms. The uncultivated or wild variety of the species is called a cardoon. It is a perennial plant native to the Mediterranean region.

Description of the Plant

It grows to 1.4–2 m (4.6–6.6 ft) tall, with arching, deeply lobed, silvery, glaucous-green leaves 50–82 cm (20–32 in) long. The flowers develop in a large head from an edible bud about 8–15 cm (3.1–5.9 in) diameter with numerous triangular scales; the individual florets are purple. The edible portions of the buds consist primarily of the fleshy lower portions of the involucral bracts and the base, known as the "heart"; the mass of immature florets in the centre of the bud is called the "choke" or beard. These are inedible in older, larger flowers.

Early History of Use

The naturally occurring variant of the artichoke, the cardoon (*Cynara cardunculus*), which is native to the Mediterranean area, has records of use as a food among the ancient Greeks and Romans. In North Africa, where they are still found in the wild state, the seeds of artichokes, probably cultivated, were found during the excavation of Roman-period Mons Claudianus in Egypt. Varieties of artichokes were cultivated in Sicily since the classical period of the ancient Greeks, the Greeks calling them *kaktos*. In that period the Greeks ate the leaves and flower heads, which cultivation had already improved from the wild form.

The Romans called the vegetable *carduus* (whence the name cardoon). Globe artichokes are known to have been cultivated at Naples around the middle of the 9th century. Further improvement in the cultivated form appears to have taken place in the medieval period in Muslim Spain and the Maghreb, although the evidence is inferential only. Names for the artichoke in many European languages today come from medieval Arabic ÇáÎÑÔæÝ *al-khurshûf* via late medieval Spain.

Le Roy Ladurie, in his book *Les Paysans de Languedoc*, has documented the spread of artichoke cultivation in Italy and southern France in the late 15th and early 16th centuries, when the artichoke appears as a new arrival with a new name, which may be taken to indicate an arrival of an improved cultivated variety:

"The blossom of the thistle, improved by the Arabs, passed from Naples to Florence in 1466, carried by Filippo Strozzi. Towards 1480 it is noticed in Venice, as a curiosity. But very soon veers towards the northwest...Artichoke beds are mentioned in Avignon by the notaries from 1532 onward; from the principle [*sic*] towns they spread into the hinterlands...appearing as *carchofas* at Cavaillon in 1541, at Chateauneuf du Pape in 1553, at Orange in 1554. The local name remains *carchofas*, from the Italian *carciofo*...They are very small, the size of a hen's egg...and are still considered a luxury, a vaguely aphrodisiac tidbit that one preserved in sugar syrup."

The Dutch introduced artichokes to England, where they grew in Henry VIII's garden at Newhall in 1530. They were brought to the United States in the 19th century, to Louisiana by French immigrants and to California by Spanish immigrants.

Agricultural Output

Today, globe artichoke cultivation is concentrated in the countries bordering the Mediterranean basin. The main European producers are Italy, Spain, and France. In the United States, California provides nearly 100% of the U.S. crop, and about 80% of that is grown in Monterey County; there, Castroville proclaims itself to be "The Artichoke Centre of the World", and holds the annual Castroville Artichoke Festival. Most recently, artichokes have been grown in South Africa in a small town called Parys located along the Vaal River. According to FAO, the top 10 artichoke producing countries are [in metric tonnes (2010)]:

1. Italy 480,112
2. Egypt 215,514
3. Spain 166,700
4. Peru 127,323
5. Argentina 84,000
6. China 70,000
7. Morocco 45,460
8. France 42,153
9. United States of America 39,190
10. Chile 35,000

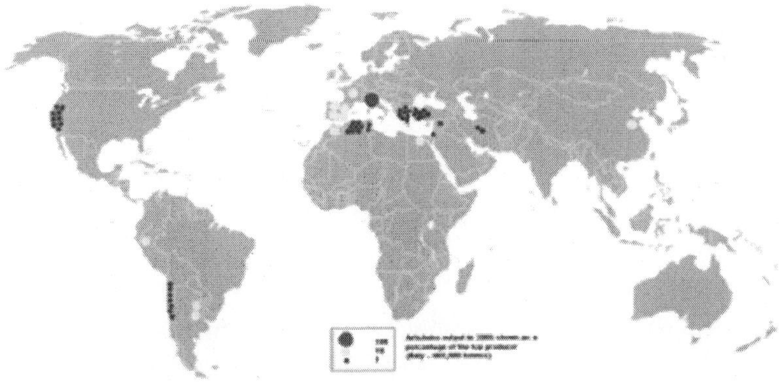

***Figure:** Artichoke output in 2005*

Artichokes can be produced from seeds or from vegetative means such as division, root cuttings or micropropagation. Though technically perennials that normally produce the edible flower only during the second and subsequent years, certain varieties of artichoke can be grown from seed as annuals, producing a limited harvest at the end of the first growing season, even in regions where the plants are not normally winter-hardy. This means home gardeners in northern regions can attempt to produce a crop without the need to overwinter plants with special treatment or protection. The recently introduced seed cultivar 'Imperial Star' has been bred to produce in the first year without such measures. An even newer cultivar, 'Northern Star', is said to be able to overwinter in more northerly climates, and readily survives subzero temperatures.

Figure: *Artichoke head with flower in bloom*

Commercial culture is limited to warm areas in USDA hardiness zone 7 and above. It requires good soil, regular watering and feeding, plus frost protection in winter. Rooted suckers can be planted each year, so mature specimens can be disposed of after a few years, as each individual plant lives only a few years. The peak season for artichoke harvesting is the spring, but they can continue to be harvested throughout the summer, with another peak period in midautumn.

When harvested, they are cut from the plant so as to leave an inch or two of stem. Artichokes possess good keeping qualities, frequently remaining quite fresh for two weeks or longer under average retail conditions. Apart from food use, the globe artichoke is also an attractive plant for its bright floral display, sometimes grown in herbaceous borders for its bold foliage and large purple flower heads.

Bean Anthracnose

Colletotrichum Lindemuthianum

Anthracnose of bean is distributed worldwide, but is more serious in temperate and subtropical zones. The pathogen causes one of the most important diseases of legume crops and yield losses can reach 100% when contaminated seeds are planted and environmental conditions remain favourable for disease development for long periods.

Identification: The disease affects all aerial parts of the bean plant. Diseased cotyledons exhibit small, dark brown to black lesions. The developing hypocotyl can become infected, developing rust coloured specks which enlarge longitudinally resulting in sunken lesions. Lesions are more common on leaf petioles, the lower surface of leaves, and leaf veins where they are elongate, angular, and brick-red to purple becoming black. Pod infections occur as tan to rust coloured lesions that become sunker cankers with a raised black ring and brownish red border.

Life Cycle: Bean anthracnose survives between crops in infected crop residue and can be disseminated in seeds, air, and water. Lesions begin producing fruiting bodies (acervuli) and conidia within 2-4 days of infection. Production of conidia and new plant infections are favoured by temperatures of 13-26° C and relative humidity greater than 92 % (or free moisture). Frequent, moderate rainfall, particularly if accompanied by winds or splashing rain favour disease development.

Crop Injury: Infected hypocotyls may rot and collapse, resulting in a poor stand. Severely affected pods may shrivel and dry. The fungus can invade the pods and infect the seed coat and cotyledons of developing seed.

Cultural Controls & Prevention:

- Start with certified, disease free seed. Production of pathogen-free seeds in dry areas has reduced disease incidence and severity.
- Certain cultivars of bean are resistant to anthracnose.
- Rotate with non-host crops such as cereals and corn, for 2-3 years.
- Promptly incorporate crop debris after harvest to hasten decomposition.
- Treat seed with appropriate fungicides to control seed coat infections.

Bean, Lima, Downy Mildew

Phytophthora Phaesoli

Downy Mildew caused by *Phytophthora phaseoli* infects broad bean, pea, and lima bean (*Phaseolus lunatus*). The disease can be a major problem in cool, moist regions of Europe and the United States.

The pathogen is adaptable and continues to produce new races (strains) on previously resistant cultivars. There may be physiological differences which limit cross infection among the various races and hosts.

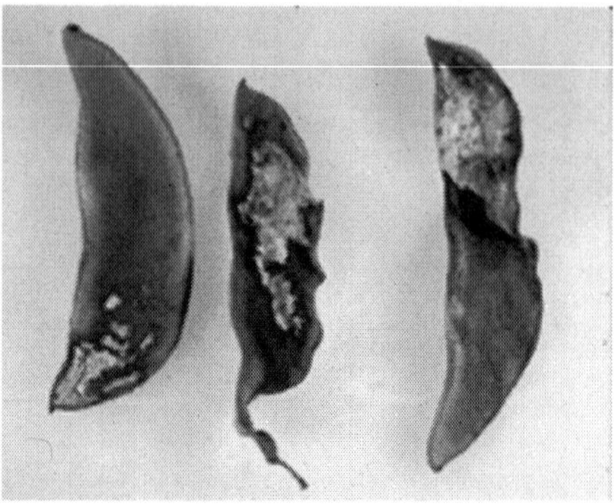

Identification: Local lesions begin as small, yellow blotches on the upper sides of leaves which are limited by leaf veins. The undersides of these lesions produce a gray to purple, fuzzy sporulation. Lesions become dry and brown and may be invaded by secondary organisms like *Botrytis*. Pods can also be attacked and exhibit large, yellow blotches on the pods and white mycelial growth within the pods.

Life Cycle: Seedlings may be infected at emergence and die, although infected seeds usually fail to germinate. Systemically infected plants are light green with a silvery appearance. *P. viciae* is seedborne and soilborne. Disease development is most rapid under cool, wet conditions. The disease is disseminated by wind and rain.

Cultural Controls & Prevention:

- Rotate with non-host crops to prevent the buildup of soilborne inoculum.
- Plow under infected crop residue after harvest.
- Plant resistant cultivars where available. Resistance may be overcome by new races of the pathogen
- Use seed treatments to protect emerging seedlings from infection.

Limo Bean

Phaseolus lunatus is a legume. It is grown for its edible seed. It is commonly known as the butter bean or lima bean.

Origin and Uses

Phaseolus lunatus is of Andean and Mesoamerican origin. Two separate domestication events are believed to have occurred. The first, taking place in the Andes around 2000 BC, produced a large-seeded variety (Lima type), while the second, taking place in Mesoamerica around AD 800, produced a small-seeded variety (Sieva type). By around 1300, cultivation had spread north of the Rio Grande, and in the 1500s, the plant began to be cultivated in the Old World.

The small-seeded wild form (Sieva type) is found distributed from Mexico to Argentina, generally below 1,600 metres (5,200 ft) above sea level, while the large-seeded wild form (Lima type) is found distributed in the north of Peru, from 320 to 2,030 metres (1,050 to 6,660 ft) above sea level.

The Moche Culture (1-800 AD) cultivated all of the lima beans and often depicted them in their art. During the Spanish Viceroyalty of Peru, lima beans were exported to the rest of the Americas and Europe, and since the boxes of such goods had their place of origin labelled "Lima – Peru", the beans got named as such.

The term "butter bean" is widely utilised for a large, flat and yellow/white variety of lima bean (*P. lunatus* var. *macrocarpus*, or *P. limensis*). In the Southern United States the Sieva type are traditionally called butter beans, also otherwise known as the Dixie or Henderson type. In that area, lima beans and butter beans are seen as two distinct types of beans. In Spain it is called "garrofón" and constitutes one of the main ingredients of the famous Valencian Paella.

In the United Kingdom, "butter beans" refers to either dried beans which can be purchased to re-hydrate, or the canned variety which are ready to use. In culinary use, lima beans and butter beans are distinct, the former being large and yellow, the latter small and green. In areas where both are considered to be lima beans, the green variety may be labelled as "baby" (and less commonly "junior") limas.

Varieties

Both bush and pole (vine) varieties exist, the latter ranges from one to four metres in height. The bush varieties mature earlier than the pole varieties. The pods are up to 15 cm long. The mature seeds are 1 to 3 cm long and oval to kidney shaped. In most varieties the seeds are quite flat, but in the "potato" varieties the shape approaches spherical. White seeds are common, but black, red, orange and variously mottled seeds are also known. The immature seeds are uniformly green. Lima beans typically yield 2900 to 5000 kilograms of seed and 3000 to 8000 kilograms of biomass per hectare.

The seeds of the varieties listed below are white unless otherwise noted.

Bush Types

- Henderson/Thorogreen, 65 days (heirloom)
- Eastland, 68 days
- Jackson Wonder, 68 days (heirloom, seeds brown mottled with purple)
- Fordhook 242, 75 days, 1945 AAS winner

Pole Types

- Giant Speckled/Christmas/Speckled Calico, 78 days (heirloom, seeds white mottled with red)
- Big 6/Big Mama, 80 days
- King of the Garden, 85 days (heirloom)

Nutritional Value

Lima beans, like many other legumes, are a good source of dietary fibre, and a virtually fat-free source of high quality protein.

Lima beans contain both soluble fibre, which helps regulate blood sugar levels and lowers cholesterol, and insoluble fibre, which aids in the prevention of constipation, digestive disorders, irritable bowel syndrome and diverticulitis.

Blood Sugar

The high fiber content in Lima beans prevents blood sugar levels from rising too rapidly after eating them. This is due to the presence of large amounts of absorption-slowing compounds in the beans, and their high soluble fiber content. Soluble fiber absorbs water in the stomach forming a gel that slows down the absorption of the bean's carbohydrates. They can therefore help balance blood sugar levels while providing steady, slow-burning energy, which makes them a good choice for people with diabetes suffering with insulin resistance.

Heart

Soluble fiber binds with the bile acids that form cholesterol and, because it is not absorbed by the intestines, it exits the body taking the bile acids with it. As a result, the cholesterol level is lowered. They may therefore help to prevent heart disease, and may reduce the medical dosage required to combat cholesterol in the form of natural food.

Lima beans also provide folate and magnesium. Folate lowers levels of homocysteine, an amino acid that is an intermediate product in an important metabolic process called the methylation cycle. Elevated blood levels of homocysteine are an independent risk factor for heart attack, stroke, and peripheral vascular disease.

The magnesium content of lima beans is a calcium channel blocker. When enough magnesium is present veins and arteries relax, which reduces resistance and improves the flow of blood, oxygen and nutrients throughout the body.

Other Minerals

Apart from providing slow-burning complex carbohydrates, lima beans can increase energy levels by helping to restore more iron. For menstruating women, who are more at risk of iron deficiency, lima beans can provide iron, an integral component of hemoglobin—hemoglobin transports oxygen from the lungs to all body cells, and is also part of key enzyme systems for energy production and metabolism.

Lima beans are a very good source of the trace mineral manganese, and help enzymes important for energy production and antioxidant defence. Lima beans also contain the trace mineral molybdenum, an integral component of the valuable enzyme sulfite oxidase; sulfite oxidase detoxifies sulfites found as food preservatives.

Raw lima beans and butter beans contain toxic linamarin, a cyanogenic glucoside—one handful of raw beans can make a person violently ill. The beans are rendered safe when cooked. Low-linamarin varieties are typically used for culinary purposes.

Insects and Pests: Lima beans are susceptible to a variety of insects, most notably beetles. They can be effectively treated with Sevin, or a variety of other insecticides.

Bunnies love beans! Rabbits eat the tender new leaves. If there are rabbits in your area, a rabbit fence is not a nicety, it is a necessity. They will devastate a row of beans in a hurry, eating the tender new leaves. As new ones develop, they will come back for more.

Plant Disease: Lima Bean plants seem to bean little less susceptible to plant disease than other beans. But, it can still happen, especially in wet, humid weather.

Bacterial and wilt diseases are common among the Bean family. This plant disease arrives with summer heat and humidity. This often occurs just before, or during, the ripening of the crop. Fungicides are recommended in areas of high heat and humidity.

Tip: Keep the leaves dry and allow more spacing between plants for better air circulation.

Hardiness: Bean plants are not hardy. They are susceptible to cold and frost. Hold off planting until a few days before all danger of frost is past. In the fall, cover the crop on nights when the temperature is expected to go below 40 degrees.

Beet

The beetroot, also known in North America as the table beet, garden beet, red or golden beet, or informally simply as the beet, refers to any of the cultivated varieties of beet (*Beta vulgaris*) grown for their edible taproots, especially *B. vulgaris* L. subsp. *conditiva.* They are among the most commonly encountered varieties in North America, Central America, and Europe.

Food

The usually deep red roots of beetroot are eaten either grilled, boiled, or roasted as a cooked vegetable, cold as a salad after cooking and adding oil and vinegar, or raw and shredded, either alone or combined with any salad vegetable. A large proportion of the commercial production is processed into boiled and sterilised beets or into pickles. In Eastern Europe, beet soup, such as borscht, is a popular dish. In Indian cuisine, chopped, cooked, spiced beet is a common side dish. Yellow-coloured beetroots are grown on a very small scale for home consumption.

The green, leafy portion of the beet is also edible. It is most commonly served boiled or steamed, in which case it has a taste and texture similar to spinach. Those selected should be bulbs that are unmarked, avoiding those with overly limp leaves or wrinkled skins, both of which are signs of dehydration. Beetroot can be peeled, steamed, and then eaten warm with butter as a delicacy; cooked, pickled, and then eaten cold as a condiment; or peeled, shredded raw, and then eaten as a salad. Pickled beets are a traditional food of the American South. It is also common in Australia, New Zealand, and the United Arab Emirates for pickled beetroot to be served on a hamburger.

A traditional Pennsylvania Dutch dish is red beet eggs. Hard-boiled eggs are refrigerated in the liquid left over from pickling beets and allowed to marinate until the eggs turn a deep pink-red colour.

In Poland, beet is combined with horseradish to form popular Æwik[3]a z chrzanem, which is often added to a meal consisting of meat, potatoes and a salad.

When beet juice is used, it is most stable in foods with a low water activity, such as frozen novelties and fruit fillings. Betanins, obtained from the roots, are used industrially as red food colorants, e.g. to intensify the colour of tomato paste, sauces, desserts, jams and jellies, ice cream, sweets and breakfast cereals.

Beet pulp is fed to horses that are in vigorous training or conditioning and to those that may be allergic to dust from hay.

Beetroot can also be used to make wine.

The consumption of beets causes pink urine in some people.

Beetroot juice has been found to improve performance in athletes, possibly because of its abundance of nitrate.

Nutrients and Phytochemicals

Beetroot is an excellent source of folate and a good source of manganese, and contains betaines which may function to reduce the concentration of homocysteine, a homolog of the naturally occurring amino acid cysteine. High circulating levels of homocysteine may be harmful to blood vessels and thus contribute to the development of heart disease, stroke, or peripheral vascular disease. This hypothesis is controversial as it has not been established yet whether homocysteine itself is harmful, or whether it is just an indicator of increased risk for heart disease.

The red colour compound betanin is not broken down in the body, and in higher concentration may temporarily cause urine (termed beeturia) and stool to assume a reddish colour. This effect may cause distress and concern due to the visual similarity to hematuria (blood in the urine) or blood in the stool, but is completely harmless and will subside once the food is out of the system.

A recent study highlighted beetroot as a source of acute dietary nitrate, which was used to test the influence of nitrate supplementation on resting heart rate and sustained apnea. 70 ml of beetroot juice, containing approximately 5mmol of nitrate, was found to reduce resting blood pressure by 2% and increase the maximum duration of apnea by 11% in experienced divers, relative to a control group receiving a placebo containing 0.003mmol nitrate.

Preliminary Research

Basic research on both rats and pilot studies on humans have shown betaine may protect against liver disease, particularly the buildup of fatty deposits in the liver caused by alcohol abuse, protein deficiency, or diabetes, among other causes.

In preliminary research, beetroot juice lowered blood pressure and thus may help reduce the risk of cardiovascular disease. The effect is attributed to the high nitrate content of the beetroot. The study correlated high nitrate concentrations in the blood following ingestion of the beetroot juice and the drop in blood pressure. Dietary nitrate, such as that found in the beetroot, is thought to be a source for the biological messenger nitric oxide, which is used by the endothelium to signal smooth muscle, triggering it to relax. This induces vasodilation and increased blood flow.

One study found positive effects beetroot may have on human exercise performance, showing that distance runners ran 5% faster times after consuming baked beetroot.

Other Uses

Betanin, obtained from the roots, is used industrially as red food colorants, e.g. to improve the colour and flavour of tomato paste, sauces, desserts, jams and jellies, ice cream, sweets, and breakfast cereals. Beetroot dye may also be used in ink.

Within older bulbs of beetroot, the colour is a deep crimson, and the flesh is much softer.

It is a rich source of the element boron.

Historical Facts

Field Marshal Montgomery is reputed to have exhorted his troops to 'take favours in the beetroot fields', a euphemism for visiting prostitutes.

From the Middle Ages, beetroot was used as a treatment for a variety of conditions, especially illnesses relating to digestion and the blood. Bartolomeo Platina recommended taking beetroot with garlic to nullify the effects of 'garlic-breath'.

Varieties

Below is a list of several commonly available varieties of beets. Generally 55 to 65 days from germination to harvest of the root. All varieties can be harvested earlier for use as greens. Unless otherwise noted, the root colour of the following varieties are shades of red and dark red with different degrees of zoning noticeable in slices.

- Albino, heirloom (white root)
- Bull's Blood, heirloom
- Golden Beet / Burpee's Golden, heirloom (yellow root)
- Chioggia, heirloom (distinct red and white zoned root)
- Cylindra / Formanova, heirloom (elongated root)

- Detroit Dark Red Medium Top, heirloom
- Early Wonder, heirloom
- Perfected Detroit, 1934 AAS winner
- Red Ace Hybrid
- Ruby Queen, 1957 AAS winner
- Touchstone Gold (yellow root)

Beetroot Pests, Diseases and Problems

Beetroot is one of those wonderful crops that tends to be relatively trouble free. Birds may have a go at the seedlings, in which case you can cover them with horticultural fleece until they are larger and less appealing to our feathered friends.

Bolting may be a problem. This is where the plant flowers before the roots have formed. This can occur in dry conditions, or if thinning has been left too late. If you have trouble with bolting, then try growing a bolt resistant variety such as Bolthardy.

Aphids may attack the leaves causing them to curl over. We find squashing them between your fingers the most effective way of getting rid of them, particularly if there are not too many. Otherwise, you can spray the leaves with horticultural soap, or a weak solution of washing up liquid such as Ecover.

Fungal Leaf Spot sometimes affects beetroot. This appears as round brown spots on the leaves. The middle of the spots sometimes fall out to leave holes. This is not usually very serious. Just remove any badly affected leaves, and remove all fallen leaves at the end of the season to reduce the number of overwintering fungal spores.

Carrot

The carrot (*Daucus carota* subsp. *sativus*; etymology: from Late Latin *carôta*, from Greek êáñüôïí *karôton*, originally from the Indo-European root *ker*- (horn), due to its horn-like shape) is a root vegetable, usually orange in colour, though purple, red, white, and yellow varieties exist. It has a crisp texture when fresh. The most commonly eaten part of a carrot is a taproot, although the greens are sometimes eaten as well. It is a domesticated form of the wild carrot *Daucus carota*, native to Europe and southwestern Asia.

The domestic carrot has been selectively bred for its greatly enlarged and more palatable, less woody-textured edible taproot. The Food and Agriculture Organisation of the United Nations (FAO) reports that world production of carrots and turnips (these plants are combined by the FAO for reporting purposes) for calendar year 2011 was almost 35.658 million tonnes. Almost half were grown in China. Carrots are widely used in many cuisines, especially in the preparation of salads, and carrot salads are a traditional in many regional cuisines.

Description

The carrot is a variable biennial plant, usually growing up to 1 m tall and flowering from June to August. The umbels are claret-coloured or pale pink before they open, then bright white and rounded when in full flower, measuring 3–7 cm wide with a festoon of bracts beneath; finally, as they turn to seed, they contract and become concave like a bird's nest. The dried umbels detach from the plant, becoming tumbleweeds.

Similar in appearance to the deadly poison hemlock, *D. carota* is distinguished by a mix of bi-pinnate and tri-pinnate leaves, fine hairs on its stems and leaves, a root that smells like carrots, and occasionally a single dark red flower in its centre.

History

The wild ancestors of the carrot are likely to have come from Iran and Afghanistan, which remain the centre of diversity of *Daucus carota*, the wild carrot. Selective breeding over the centuries of a naturally occurring subspecies of the wild carrot, *Daucus carota* subsp. *sativus*, to reduce bitterness, increase sweetness and minimise the woody core, has produced the familiar garden vegetable.

In early use, carrots were grown for their aromatic leaves and seeds, not their roots. Carrot seeds have been found in Switzerland and Southern Germany dating to 2000–3000 BC. Some relatives of the carrot are still grown for their leaves and seeds, such as parsley, fennel, dill and cumin. The first mention of the root in classical sources is in the 1st century.

The modern carrot originated in Afghanistan about 1100 years ago. It appears to have been introduced to Europe via Spain by the Moors in the 8th century.

The 12th-century Arab Andalusian agriculturist, Ibn al-'Awwam, describes both red and yellow carrots; Simeon Seth also mentions both colours in the 11th century. Cultivated carrots appeared in China in the 14th century, and in Japan in the 18th century.

Orange-coloured carrots appeared in the Netherlands in the 17th century. These, the modern carrots, were intended by the antiquary

John Aubrey (1626–1697) when he noted in his memoranda "Carrots were first sown at Beckington in Somersetshire Some very old Man there [in 1668] did remember their first bringing hither." European settlers introduced the carrot to Colonial America in the 17th century.

Chemistry

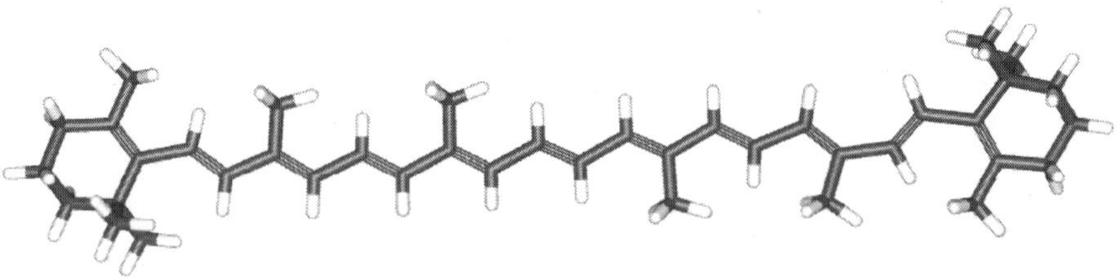

Figure: *β-Carotene structure. Carotene is responsible for the orange colour of carrots and many other fruits and vegetables.*

Polyacetylenes can be found in Apiaceae vegetables like carrots where they show cytotoxic activities. Falcarinol and falcarindiol (*cis*-heptadeca-1,9-diene-4,6-diyne-3,8-diol) are such compounds. This latter compound shows antifungal activity towards *Mycocentrospora acerina* and *Cladosporium cladosporioides*. Falcarindiol is the main compound responsible for bitterness in carrots.

Other compounds such as pyrrolidine (present in the leaves), 6-hydroxymellein, 6-methoxymellein, eugenin, 2,4,5-trime-thoxy-benzaldehyde (gazarin) or (Z)-3-acetoxy-heptadeca-1,9-diene-4,6-diin-8-ol (falcarindiol 3-acetate) can also be found in carrot.

Nutrition

Most carrot cultivars are about 88% water, 7% sugar, 1% protein, 1% fibre, 1% ash, and 0.2% fat. The fibre comprises mostly cellulose, with smaller proportions of hemicellulose and lignin. Carrots contain almost no starch. Free sugars in carrot include sucrose, glucose, xylose and fructose. Nitrite and nitrate contents are about 40 and 0.41 milligrams per 100 grams (fresh), respectively. Most of the taste of the vegetable is due to glutamic acid and other free amino acids. Other acids present in trace amounts include succinic acid, α-ketoglutaric acid, lactic acid and glycolic acid; the major phenolic acid is caffeic acid.

The carrot gets its characteristic and bright orange colour from β-carotene, and lesser amounts of α-carotene and γ-carotene. α and β-carotenes are partly metabolised into vitamin A in humans. β-carotene is the predominant carotenoid, although there are lesser amounts of α-carotene and γ-carotene. There are typically between

6000 and 54,000 micrograms of carotenoids per 100 grams of carrot root. Carrot extracts are used by poultry producers to improve animal skin and alter the colour of egg yolk. Massive overconsumption of carrots can cause carotenosis, a benign condition in which the skin turns orange. Carrots are also rich in antioxidants and minerals. Ethnomedically, the roots are used to as an emmenagogue (to increase blood flow in the pelvic area and uterus), a carminative (to reduce flatulence), to treat digestive problems, intestinal parasites, and tonsillitis or constipation.

Lack of vitamin A can cause poor vision, including night vision, and these can be restored by adding vitamin A to the diet. An urban legend states that eating large quantities of carrots will allow one to see in the dark. This myth developed from stories about British gunners in World War II, who were able to shoot down German planes at night. The rumour arose during the Battle of Britain when the RAF circulated a story about their pilots' carrot consumption in an attempt to cover up the discovery and effective use of radar technologies in engaging enemy planes, as well as the use of red light (which does not destroy night vision) in aircraft instruments. It reinforced existing German beliefs, and helped to encourage Britons who were trying to improve their night vision during the blackout to grow and eat the vegetable, which was not rationed like most other foodstuffs. A "Dr. Carrot" advertising campaign encouraged its consumption.

Methods of Consumption and Uses

Carrots can be eaten in a variety of ways. Only 3% of the â-carotene in raw carrots is released during digestion: this can be improved to 39% by pulping, cooking and adding cooking oil. Alternatively they may be chopped and boiled, fried or steamed, and cooked in soups and stews, as well as baby and pet foods. A well-known dish is *carrots julienne*. Together with onion and celery, carrots are one of the primary vegetables used in a *mirepoix* to make various broths.

The greens are edible as a leaf vegetable, but are only occasionally eaten by humans; some sources suggest that the greens contain toxic alkaloids. When used for this purpose, they are harvested young in high-density plantings, before significant root development, and typically used stir-fried, or in salads. Some people are allergic to carrots. In a 2010 study on the prevalence of food allergies in Europe, 3.6 percent of young adults showed some degree of sensitivity to carrots. Because the major carrot allergen, the protein Dauc c 1.0104, is cross-reactive with homologues in birch pollen (Bet v 1) and mugwort pollen (Art v 1), most carrot allergy sufferers are also allergic to pollen from these plants. In India carrots are used in a variety of ways, as salads or as vegetables added to spicy rice or daal dishes. A popular variation in north India is the Gajar Ka Halwa carrot dessert, which has carrots grated and cooked in milk until the whole mixture is solid, after which nuts and butter are added. Carrot salads are usually made with grated carrots in western parts with a seasoning of mustard seeds and green chillies popped in hot oil, while adding carrots to rice usually is in julienne shape.

The variety of carrot found in north India is rare everywhere except in Central Asia and other contiguous regions, and is now growing in popularity in larger cosmopolitan cities in South India. The north Indian carrot is pink-red comparable to plum or raspberry or deep red apple in colour (without a touch of yellow or blue) while most other carrot varieties in the world vary from orange to yellow in colour, comparable to hallowe'en pumpkins.

Since the late 1980s, baby carrots or mini-carrots (carrots that have been peeled and cut into uniform cylinders) have been a popular ready-to-eat snack food available in many supermarkets. Carrots are puréed and used as baby food, dehydrated to make chips, flakes, and powder, and thinly sliced and deep-fried, like potato chips.

The sweetness of carrots allows the vegetable to be used in some fruit-like roles. Grated carrots are used in carrot cakes, as well as carrot puddings, an old English dish thought to have originated in the early 19th century. Carrots can also be used alone or with fruits in jam and preserves. Carrot juice is also widely marketed, especially as a health drink, either stand-alone or blended with fruits and other vegetables.

6

Pests and Diseases

The three most important insect pests of carrots in Ontario are: (a) the carrot rust fly, *Psila rosae*, (Fab.); (b) the carrot weevil, *Listronotus oregonensis* (Le Conte); and (c) the aster (formerly six-spotted) leafhopper, *Macrosteles quadrilineatus* (Forbes). Larvae of the carrot rust fly and carrot weevil create tunnels in the roots of carrots, making them unmarketable. The aster leafhopper carries and transmits the mycoplasma-like organism causing aster yellows (AY-MLO), a disease that affects carrots as well as a wide range of crops.

These pests are of most concern in the Holland and neighbouring marshes, where more than 80 per cent of Ontario's carrots are grown. All three pests are present to varying degrees in much of southern Ontario. Timing of seasonal events and importance of pest damage varies in the province according to climate and soil type. For example, in south-western Ontario, the aster leafhopper-aster yellows complex is the major pest.

Identifying Damage

Although carrot weevil and carrot rust fly damage may appear similar in the early stages, normally they are distinguished easily on mature carrots. Carrot weevil feeding tunnels tend to occur in the upper one-third of the root, with a conspicuous, darkened, partly open tunnel in the crown after the larva has matured and left the carrot. The larva is a legless grub with a reddish-brown head.

Carrot rust fly tunnels, found mainly in the lower two-thirds of the root, are narrower and more winding than those of the carrot weevil. The larvae are legless and headless maggots. In severe early-season attacks by either insect, some young plants may wilt or be killed, although that type of damage generally is not economically important in commercial fields. The nature of first-generation carrot rust fly injury depends upon the development of the plant at the time of the attack. There may be tunnelling or destruction of the extremity of the tap root only, or, if the root has thickened, tunnels similar to those caused by the second generation may occur. Both pests can also attack celery, parsnip, wild carrot and other Umbelliferae. In severe attacks, young celery plants can be destroyed.

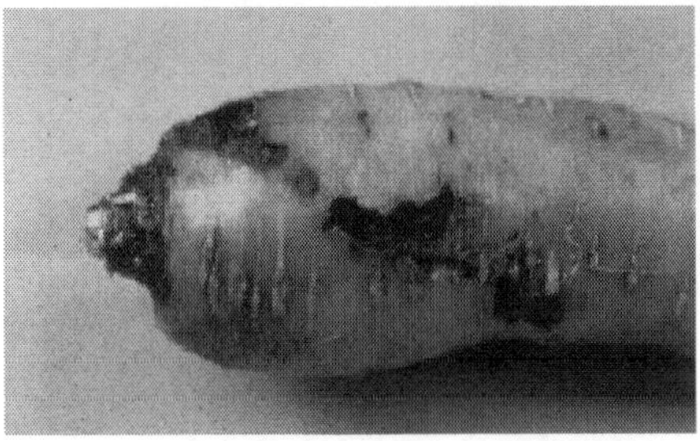

Figure: *Feeding tunnels of the carrot weevil, formed mainly in the upper one-third of the root.*

Aster leafhoppers are inconspicuous and evidence of their attack shows up much later in the form of aster yellows. Infected carrots have excessive yellow, twisted, fernlike foliage, with hairy roots

Carrot Rust Fly

Life Cycle and Habits

The adult carrot rust fly is a slender, shiny, black fly, about 6 mm long, with a small but characteristic reddish head and long yellow legs. The insect overwinters in the soil in the pupal stage in a small, seedlike puparium.

Figure: *Aster yellows of carrot, with excessive yellow, twisted foliage and hairy roots.*

Adults begin to emerge in late May, after about 250 degree-days above 5 °C, and may be present until near the end of June. They lay eggs in the soil around young carrots. The young larvae feed at first on root hairs and small rootlets and after their third moult, enter the carrots to complete their development. When mature, they pupate in the soil. Near the end of July, after about 1150 degree-days above 5 °C, another generation of adults emerges that may last for nearly two months. The larvae of the second-generation pupae spend the winter in the soil, although in some years a few adults may emerge in October. Normally it is too late for them to cause any extensive damage.

Management

Carrot rust fly adults can be monitored with orange/yellow sticky traps placed in the carrot field or in nearby trees where the adults rest.

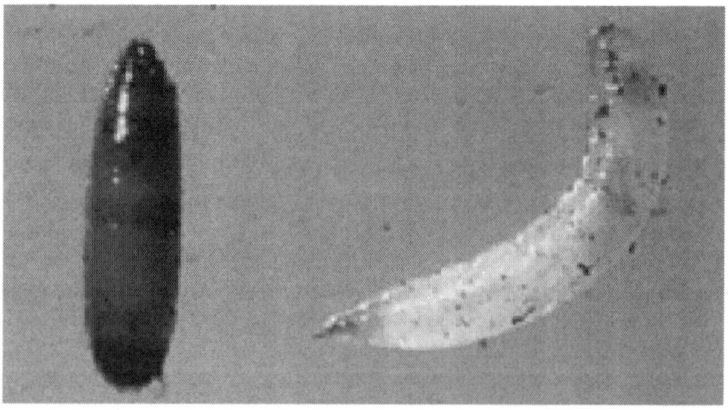

Figure: *Pupa of carrot rust fly, left; larva, right.*

Such monitoring indicates if the rust fly is present at levels high enough to warrant an insecticide application. Treatments can be properly timed based on monitoring information. Consult your local Horticultural Crops Advisor for specific recommendations as well as OMAFRA Publication 363, *Vegetable Production Recommendations.*

Also consult OMAFRA publication, *Integrated Pest Management for Onions, Carrots, Celery and Lettuce in Ontario*, Order No. 700, for monitoring thresholds and treatment guidelines. Home gardeners may consult the most recent issue of OMAFRA publication 64, *Insect and Disease Control in the Home Garden.* Insecticides are rarely recommended for small gardens.

Monitoring of individual fields or farms could enable some growers to reduce the number of sprays applied. Because of the shelter-seeking habits of the flies, fields located in open areas often escape with little damage and need fewer sprays than those in more sheltered locations. Similarly, isolated carrot fields in other parts of the province may be free from the insect and not require sprays.

Dates of planting and harvest also affect the need to use insecticides. In the Holland Marsh, carrots seeded after mid-May normally are not seriously injured by the first generation larvae and do not require spraying for that generation. Spraying within one month before harvest is not required, because at least one month is required after egg deposition until larvae enter the carrots.

The sheltering behaviour of the carrot rust fly often results in a gradient of damage in the field if damage is severe near the ends of the rows but declines rapidly with distance from the shelter. This can create problems when harvesting mechanically, because carrots with different levels of infestation can become mixed in the pallet boxes. By leaving badly infested areas unharvested, however, the number of overwintering pupae in the soil will be greater in the following year.

As indicated above, the period of attack by second generation larvae continues from early September until the end of harvest, during which time the number of infested carrots will increase. By examining samples of the crop periodically, beginning in mid-September, an indication of the ultimate level of carrot rust fly injury can be obtained. The crop could then be harvested earlier to minimize the proportion of damaged carrots.

Carrot Weevil

Life Cycle and Habits

The carrot weevil adult is a dark-brown snout beetle about 6 mm long. It overwinters in plant debris in and about carrot fields that were infested the previous year.

Figure: *Adult carrot weevil.*

In spring, adults become active and mate after a few warm days. They are capable of laying eggs by mid- to late May. However, they do not attack the new crop of carrots until the first true leaf stage. Adult females chew small cavities in the petioles or crown of the carrots and deposit an average of two to three eggs in each, sealing the cavity with a black exudate.

Eggs hatch after one to two weeks and the young larvae tunnel down into the root or leave the stalk and enter the roots from the soil. Some young plants may wilt and die as the slender root is tunnelled by the developing larva. The damage is not otherwise conspicuous until the larvae are nearly mature.

After feeding for at least three weeks, larvae leave the carrot and pupate in the soil. After one to two weeks, adults emerge. If warm weather or an early carrot crop has permitted adults to mature early enough in the summer, some second-generation eggs and larvae may occur.

Normally, however, there is only one generation in the Holland Marsh. In warmer areas of the continent, second-generation injury is more common. Behaviour of the summer adults after they emerge is not well understood, but they do overwinter as described above.

Management

Monitoring the activity of carrot weevil adults is an effective means of determining the level of infestation and forecasting the need for insecticide treatments. Wooden-plate traps or carrot-root sections placed in the soil, can be used to monitor adult activity, starting at the time of seeding.

Consult OMAF publication, *Integrated Pest Management for Onions, Carrots, Celery and Lettuce in Ontario*, Order No. 700, for monitoring and threshold guidelines. Treatment thresholds are based on a cumulative weevil per trap count. Between zero and two applications may be necessary, depending on the level of weevil activity in the field. In some cases, border applications may provide adequate control of adults. Parasitism by an egg parasitoid, Anaphes sp. is a significant control factor and should be encouraged. Border areas with weeds offer a suitable habitat for the parasitoid. Carrot weevil adults rarely fly and therefore the insect does not spread rapidly. Its presence in a particular location should be evident for a season or two before it builds up to serious levels. Development of a serious infestation can be delayed by not planting carrots on or adjacent to sites that were infested the preceding year.

Aster Leafhopper

Life Cycle and Habits: Aster leafhopper adults are inconspicuous, active insects, about 3 mm long, that feed and breed upon a wide range of cultivated and weed hosts. In Ontario, the insect overwinters in the egg stage on winter grain. The immature stages, or nymphs, hatch in early May and become adults about two weeks later. The adults move from winter grain to various hosts such as spring grain, grasses, vegetables and weeds. During the summer, two to five generations of leafhoppers develop on the various hosts, some of which will harbour AY-MLO. Adults moving to vegetables then introduce the organism and later transmit it within the crop.

In central Ontario, aster leafhoppers first appear on carrots in late May or early June and occur throughout the balance of the season. In southwestern Ontario and the Niagara Peninsula, they usually appear in early May and often in very high numbers.

Figure: *Adult aster leafhopper.*

Management

Reduction of weed hosts through the use of herbicides has reduced the number of aster yellows in the Holland and neighbouring marshes. However, in many parts of southern Ontario, the potential for disease is serious, although it varies from year to year. Leafhopper adults can be monitored with sticky traps or with a sweep net. Currently, there is no way of forecasting what level of aster yellows infectivity may be present in the leafhoppers, however, this capability should be available in the near future.

Some of the insecticides used in carrot rust fly control are also effective against leafhoppers. Therefore, special sprays for leafhopper control will not likely be required unless there is a large increase in leafhopper numbers between generations of the rust fly. In southern

Ontario, leafhopper populations may be heavy, and if sprays for carrot rust fly are not required, it may still be necessary to spray for leafhopper control.

It is important to eliminate potential wild hosts of leafhoppers or aster yellows in the vicinity of carrot fields and to remove any infected carrot plants that are detected. Among the weeds that are hosts for aster yellows are thistles, fleabane, wild lettuce, sow-thistle, chicory, wild carrot, galinsoga, dandelion, plantain, cinquefoil and others. Crops such as lettuce, endive, parsnip and celery may also be affected by aster yellows, as well as numerous florist crops in the aster family.

Cultivars

Carrot cultivars can be grouped into two broad classes, eastern carrots and western carrots. More recently, a number of novelty cultivars have been bred for particular characteristics.

The city of Holtville, California, promotes itself as "Carrot Capital of the World", and holds an annual festival devoted entirely to the carrot. Eastern carrots were domesticated in Central Asia, probably in modern-day Iran and Afghanistan in the 10th century, or possibly earlier. Specimens of the eastern carrot that survive to the present day are commonly purple or yellow, and often have branched roots. The purple colour common in these carrots comes from anthocyanin pigments.

The western carrot emerged in the Netherlands in the 17th century, its orange colour making it popular in those countries as an emblem of the House of Orange and the struggle for Dutch independence. The orange colour results from abundant carotenes in these cultivars.

Western carrot cultivars are commonly classified by their root shape:

- Chantenay carrots are shorter than other cultivars, but have greater girth, sometimes growing up to 8 centimetres (3 in) in diameter. They have broad shoulders and taper towards a blunt, rounded tip. They are most commonly diced for use in canned or prepared foods.
- Danvers carrots have a conical shape, having well-defined shoulders and tapering to a point at the tip. They are somewhat shorter than Imperator cultivars, but more tolerant of heavy soil. Danvers cultivars are often puréed as baby food. They were developed in 1871 in Danvers, Massachusetts.

One particular variety lacks the usual orange pigment from carotenes, owing its white colour to a recessive gene for tocopherol (vitamin E). Derived from *Daucus carota* L. and patented at the University of Wisconsin–Madison, the variety is intended to supplement the dietary intake of Vitamin E.

Figure: *Carrots come in a wide variety of shapes, colours and sizes.*

Production Trends

Carrot is one of the top-ten most economically important vegetables crops in the world. In 2011, according to the Food and Agriculture Organisation of the United Nations, 35.658 million tonnes of carrots and turnips were produced worldwide for human consumption, grown on 1,184,000 hectares (2,926,000 acres).

With 16.233 million tonnes, China was by far the largest producer and accounted for 45.5% of the global output, followed by Russia (1.735 million tonnes), the United States (1.342), Uzbekistan (1.222), Poland (0.887), Ukraine (0.864), and the United Kingdom (0.694). About 61% of world carrot production occurred in Asia, followed by the Europe (24.2%) and the Americas (North, Central, and South America and the Caribbean) (9.7%).

Less than 4% of the world's 2011 total was produced in Africa. Global production has increased from 21.4 million tonnes in 2000, 13.7 million tonnes in 1990, 10.4 million tonnes in 1980, and 7.85 million tonnes in 1970.

The rate of increase in the global production of carrots has been greater than the world's population growth rate, and greater than the overall increase in world vegetable production. Europe was traditionally the major centre of production, but was overtaken by Asia in 1997.

The growth in global production is largely the result of increases in production area, rather than average yield. Modest improvements in the latter can be attributed to optimised agricultural practices, the development of better cultivars (including hybrids), and increased farm mechanisation.

Storage

Carrots can be stored for several months in the refrigerator or over winter in a moist, cool place. For long term storage, unwashed carrots can be placed in a bucket between layers of sand, a 50/50 mix of sand and wood shavings, or in soil. A temperature range of 32 to 40°F (0 to 5°C) is best.

Celery

Celery (*Apium graveolens* var. *dulce*) is a plant variety in the family Apiaceae, commonly used as a vegetable. The plant grows to 1 m tall. The leaves are pinnate to bipinnate leaves with rhombic leaflets 3–6 cm long and 2–4 cm broad. The flowers are creamy-white, 2–3 mm diameter, produced in dense compound umbels. The seeds are broad ovoid to globose, 1.5–2 mm long and wide.

Etymology

First attested in English in 1664, the word "celery" derives from the French *céleri*, in turn from Italian *seleri*, the plural of *selero*, which comes from Late Latin *selinon*, the latinisation of the Greek óÝëéíïí (*selinon*), "parsley". The earliest attested form of the word is the Mycenaean Greek *se-ri-no*, written in Linear B syllabic script.

Taxonomy

Celery was described by Carl Linnaeus in Volume One of his *Species Plantarum* in 1753.

Cultivation

In North America, commercial production of celery is dominated by the varieties called Pascal celery. Gardeners can grow a range of cultivars, many of which differ little from the wild species, mainly in having stouter leaf stems. They are ranged under two classes, white and red; the white cultivars being generally the best flavoured, and the most crisp and tender. The stalks grow in tight, straight, parallel bunches, and are typically marketed fresh that way, without roots and just a little green leaf remaining.

In Europe the dominant variety of celery most commonly available in trade is Celeriac (*Apium graveolens var. rapaceum*) grown for its hypocotyl forming a large bulb (commonly but incorrectly called celery root). The leaves are used as seasoning, and the stalks find only marginal use. The wild form of celery is known as "smallage". It has a furrowed stalk with wedge-shaped leaves, the whole plant having a coarse, earthy taste, and a distinctive smell. The stalks are not usually eaten (except in soups or stews in French cuisine), but the leaves may be used in salads, and its seeds are those sold as a spice. With cultivation and blanching, the stalks lose their acidic qualities and assume the mild, sweetish, aromatic taste particular to celery as a salad plant. The plants are raised from seed, sown either in a hot bed or in the open garden according to the season of the year, and after one or two thinnings and transplantings, they are, on attaining a height of 15–20 cm, planted out in deep trenches for convenience of blanching, which is effected by earthing up to exclude light from the stems.

In the past, celery was grown as a vegetable for winter and early spring; it was perceived as a cleansing tonic, welcomed to counter the salt-sickness of a winter diet. By the 19th century, the season for celery had been extended, to last from the beginning of September to late in April.

Harvesting and Storage

Harvesting occurs when the average size of celery in a field is marketable; due to extremely uniform crop growth, fields are harvested only once. The petioles and leaves are removed and harvested; celery is packed by size and quality (determined by colour, shape, straightness and thickness of petiole, stalk and midrib length and absence of disease, cracks, splits, insect damage and rot). Under optimal conditions, celery can be stored for up to seven weeks between 0 to 2 °C (32 to 36 °F). Inner stalks may continue growing if kept at temperatures above 0 °C (32 °F). Freshly cut petioles of celery are prone to decay, which can be prevented or reduced through the use of sharp blades during processing, gentle handling, and proper sanitation.

In the past, restaurants used to store it in a container of water with powdered vegetable preservative; however, the sulfites in the preservative caused allergic reactions in some people. In 1986, the U.S. Food and Drug Administration banned the use of sulfites on fruits and vegetables intended to be eaten raw.

Uses

Celery is used around the world as a vegetable for the crisp petiole (leaf stalk). The leaves are strongly flavoured and are used less often, either as a flavouring in soups and stews or as a dried herb.

In temperate countries, celery is also grown for its seeds. Actually very small fruit, these "seeds" yield a valuable volatile oil used in the perfume and pharmaceutical industries. They also contain an organic compound called apiol. Celery seeds can be used as flavouring or spice, either as whole seeds or ground and mixed with salt, as celery salt. Celery salt can also be made from an extract of the roots, or using dried leaves. Celery salt is used as a seasoning, in cocktails (notably to enhance the flavour of Bloody Mary cocktails), on the Chicago-style hot dog, and in Old Bay Seasoning.

Figure: *Celery seed (*Apium graveolens*) essential oil in clear glass vial*

Celery, onions, and bell peppers are the "holy trinity" of Louisiana Creole and Cajun cuisine. Celery, onions, and carrots make up the French mirepoix, often used as a base for sauces and soups. Celery is a staple in many soups, such as chicken noodle soup.

Medicine

Figure: *Celery seeds*

The use of celery seed in pills for relieving pain was described by Aulus Cornelius Celsus around AD 30. Celery seeds contain a compound, 3-*n*-butylphthalide, that has been demonstrated to lower blood pressure in rats.

Bergapten in the seeds can increase photosensitivity, so the use of essential oil externally in bright sunshine should be avoided. The oil and large doses of seeds should be avoided during pregnancy, as they can act as a uterine stimulant. Seeds intended for cultivation are not suitable for eating as they are often treated with fungicides.

Cultural Depictions

Figure: *Selinunte didrachm coin bearing a* selinon *(celery) leaf, circa 515-470 BC.*

A chthonian symbol among the ancient Greeks, celery was said to have sprouted from the blood of Kadmilos, father of the Cabeiri, chthonian divinities celebrated in Samothrace, Lemnos and Thebes. The spicy odour and dark leaf colour encouraged this association with the cult of death. In classical Greece, celery leaves were used as garlands for the dead, and the wreaths of the winners at the Isthmian Games were first made of celery before being replaced by crowns made of pine. According to Pliny the Elder in Achaea, the garland worn by the winners of the sacred Nemean Games was also made of celery. The Ancient Greek colony of Selinous (Greek: Óåëéíïæò, *Selinous*), on Sicily, was named after wild parsley that grew abundantly there; Selinountian coins depicted a parsley leaf as the symbol of the city.

The name "celery" retraces the plant's route of successive adoption in European cooking, as the English "celery" (1664) is derived from the French *céleri* coming from the Lombard term, *seleri*, from the Latin *selinon*, borrowed from Greek.

Celery's surprisingly late arrival in the English kitchen is an end-product of the long tradition of seed selection needed to reduce the sap's bitterness and increase its sugars. By 1699, John Evelyn could recommend it in his *Acetaria. A Discourse of Sallets*: "Sellery, apium Italicum, (and of the Petroseline Family) was formerly a stranger with us (nor very long since in Italy) is an hot and more generous sort of Macedonian Persley or Smallage...and for its high and grateful Taste is ever plac'd in the middle of the Grand Sallet, at our Great Men's tables, and Praetors feasts, as the Grace of the whole Board".

Celery has made a surprising appearance in football folklore. Supporters of English Premier League team Chelsea and Football League team Gillingham regularly sing songs about the vegetable and are famed for throwing celery during matches. This has also given rise to the "Chelsea Cocktail", a pint of Guinness garnished with a stick of celery.

7

Plant Protection Measures

Fenugreek

Fenugreek *Trigonella foenum-graecum*) is an annual plant in the family Fabaceae. The plant has small round leaves, is cultivated worldwide as a semi-arid crop, and is a common ingredient in dishes from the Indian Subcontinent. It is known as *methi* in Marathi , Punjabi , Hindi , Urdu, Bengali and Nepali , as *menthiyam*, and *venthayam* in Tamil, "uluhaal" in Sinhala, Helba) in Arabic, *menthya*) in Kannada, uluwa in Malayalam, moshoseitaro or in Greek and *menthulu* (.F$A2A) in Telugu.

History

Zohary and Hopf note that it is not yet certain which wild strain of the genus *Trigonella* gave rise to the domesticated fenugreek but they believe it was brought into cultivation in the Near East. Charred fenugreek seeds have been recovered from Tell Halal, Iraq, (radiocarbon dating to 4000 BC) and Bronze Age levels of Lachish, as well as desiccated seeds from the tomb of Tutankhamen. Cato the Elder lists fenugreek with clover and vetch as crops grown to feed cattle.

Production

Major fenugreek-producing countries are India, Iran, Nepal, Bangladesh, Argentina, Egypt, France, Spain, Turkey and Morocco. The largest producer of fenugreek in the world is India, where the major fenugreek-producing states are Rajasthan, Gujarat, Uttarakhand, Uttar Pradesh, Madhya Pradesh, Maharashtra, Haryana, and Punjab. Rajasthan produces the lion's share of India's production, accounting for over 80% of the nation's total fenugreek output.

Cuisine: Fenugreek has three culinary uses: as a herb (dried or fresh leaves), as a spice (seeds), and as a vegetable (fresh leaves, sprouts, and microgreens). Sotolon is the chemical responsible for fenugreek's distinctive sweet smell.

The distinctive cuboid-shaped, yellow-to-amber coloured fenugreek seeds are frequently encountered in the cuisines of the Indian subcontinent. The seeds are used in the preparation of pickles, vegetable dishes, daals, and spice mixes, such as panch phoron and sambar powder. Fenugreek seeds are used both whole and in powdered form and are often roasted to reduce their bitterness and enhance their flavour.

Fenugreek is also used as a vegetable. Fresh fenugreek leaves are an ingredient in some Indian curries. The sprouted seeds and microgreens are used in salads. When harvested as microgreens, fenugreek is known as Samudra Methi in Maharashtra, especially in and around Mumbai, where it is often grown near the sea in the sandy tracts, hence the name (Samudra, which means "ocean" in Sanskrit). *Samudra Methi* is also grown in dry river beds in the Gangetic plains. When sold as a vegetable in India, the young plants are harvested with their roots still attached. Any remaining soil is washed off to extend their shelf life. They are then sold in small bundles in the markets and bazaars.

In Persian cuisine, fenugreek leaves are used and called ÔäÈáÌáå (shanbalile). It is the key ingredient and one of several greens incorporated into ghormeh sabzi and Eshkeneh, often said to be the Iranian national dishes.

Fenugreek is used in Eritrean and Ethiopian cuisine. The word for fenugreek in Amharic is *abesh* (or *abish*), and the seed is used in Ethiopia as a natural herbal medicine in the treatment of diabetes.

Yemenite Jews following the interpretation of Rabbi Salomon Isaacides, Rashi of Talmûd, believe fenugreek, which they call hilbeh, hilba, helba, or halba (çéìáä) is the Talmudic Rubia (øåáéà). They use fenugreek to produce a sauce also called hilbeh, reminiscent of curry. It is consumed daily but ceremoniously during the meal of the first and/or second night of Rosh Hashana (Jewish New Year).

Lactation

Fenugreek seeds are thought to be a galactagogue that is often used to increase milk supply in lactating women.

Medicinal

A June 2011 study at the Australian Centre for Integrative Clinical and Molecular Medicine found that men aged 25 to 52 who took a fenugreek extract twice daily for six weeks scored 25% higher on tests gauging libido levels than those who took a placebo. Also in Ancient Greek texts Pedanius Dioscorides, an ancient Greek pharmacologist, makes extensive remarks, such as, about its anti-inflammatory properties.

Seeds

Figure: *Dried fenugreek seed*

Fenugreek seed is widely used as a galactagogue (milk producing agent) by nursing mothers to increase inadequate breast milk supply. Studies have shown that fenugreek is a potent stimulator of breast milk production and its use was associated with increases in milk production. It can be found in capsule form in many health food stores.

Several human intervention trials demonstrated that the antidiabetic effects of fenugreek seeds ameliorate most metabolic symptoms associated with type-1 and type-2 diabetes in both humans and relevant animal models by reducing serum glucose and improving glucose tolerance.

Nutritional Profile

Fenugreek leaves (per 100 g of edible portion) contain the following nutrients:

- Carbohydrates: 6.0 g
- Protein: 4.4 g
- Fat: 0.9 g
- Minerals: 1.5 g
- Calcium: 395 mg

- Phosphorus: 51 mg
- Iron: 1.93 mg
- Total energy: 49 kcal

Plant Protection

Pest: Aphid: Aphid sucks the sap of tender parts of plants and affects the growth, adversely. Spraying of 0.03% solution of Dimethoate or 0.025% solution of Methyl demetone or 0.04% solution of Monocrotophos is recommended to control the aphid. If the crop is grown for green (vegetable) purpose then spray Malathion.

Diseases: i) Powdery mildew: This disease appears usually in later stage of crop and becomes serious when pod formation takes place. In this disease, white powdery patches appear on the lower and upper surface of leaves and other parts of plant. Crop should be dusted with 300 mesh Sulphur dust @ 25 kg/ha to control this disease as soon as the symptoms are noticed. Spraying of wettable Sulphur or Dinocap (Kerathan or Thiowet) can also be used to control the disease @ 20-25 g per 10 liter of water at the initial stage of this disease. If needed two more sprays should be given at an interval of 15 days after first spray.

Downy Mildew: This disease occurs during February and March. Yellow patches on the upper surface of leaves appear in the infected plants and white cottony mycelium on the lower surface of leaves. This disease can be controlled by spraying of 0.2% solution of Difoltan or any other copper fungicide.

Harvesting: Crop becomes ready for harvest in about 120-150 days. At the time of ripening/maturity, leaves and pods become yellowish and leaves start falling. Timely harvesting is very important for this crop as late harvest leads to seed losses due to pod bursting, whereas in early harvest, the grains remain immature and small. Harvesting should be done early in the morning. After harvest, plants should be dried in threshing yard and threshed by trampling under the feet of bullocks. Seeds should be separated and cleaned by winnowing.

Kohlrabi

Kohlrabi (German turnip) (*Brassica oleracea* Gongylodes group) (Olkopi in Assamese and Bengali) is a perennial vegetable, and is a low, stout cultivar of cabbage. Kohlrabi can be eaten raw as well as cooked.

Figure: *Kohlrabi stem with leaves*

The name comes from the German *Kohl* ("cabbage") plus *Rübe ~ Rabi* (Swiss German variant) ("turnip"), because the swollen stem resembles the latter, hence its Austrian name *Kohlrübe*. Kohlrabi is a very commonly eaten vegetable in German speaking countries.

In India, Kohlrabi is more commonly called *Knolkhol* (English) or *Nookal* (Hindi). It is also used extensively in the southern part of India. In Kannada, Kohlrabi is called *Gedde Kosu* or *Navilu Kosu*. In Kashmiri, the swollen stems are called *Moonji* (singular: *Muund*) and the leaves are called *Haakh* or *munji Haakh*. One commonly used variety grows without a swollen stem, having just leaves and a very thin stem, and is called *Haakh*.

Description

Kohlrabi has been created by artificial selection for lateral meristem growth (a swollen, nearly spherical shape); its origin in nature is the same as that of cabbage, broccoli, cauliflower, kale, collard greens, and Brussels sprouts: they are all bred from, and are the same species as the wild cabbage plant (*Brassica oleracea*).

The taste and texture of kohlrabi are similar to those of a broccoli stem or cabbage heart, but milder and sweeter, with a higher ratio of flesh to skin. The young stem in particular can be as crisp and juicy as an apple, although much less sweet.

Except for the *Gigante* cultivar, spring-grown kohlrabi much over 5 cm in size tend to be woody, as do full-grown kohlrabi much over perhaps 10 cm in size; the Gigante cultivar can achieve great size while remaining of good eating quality. The plant matures in 55–60 days after sowing. Approximate weight is 150 g and has good standing ability for up to 30 days after maturity.

There are several varieties commonly available, including White Vienna, Purple Vienna, Grand Duke, Gigante (also known as "Superschmelz"), Purple Danube, and White Danube. Colouration of the purple types is superficial: the edible parts are all pale yellow. The leafy greens can also be eaten.

Preparation and Use

Kohlrabi stems are surrounded by two distinct fibrous layers that do not soften appreciably when cooked. These layers are generally peeled away prior to cooking or serving raw, with the result that the stems often provide a smaller amount of food than one might assume from their intact appearance.

Kohlrabi leaves are edible and can be used interchangeably with collard and kale. Kohlrabi is an important part of the Kashmiri diet and one of the most commonly cooked foods. It is prepared with its leaves and served with a light gravy and eaten with rice.

Turnip

The turnip or white turnip (*Brassica rapa* subsp. *rapa*) is a root vegetable commonly grown in temperate climates worldwide for its white, bulbous taproot. Small, tender varieties are grown for human consumption, while larger varieties are grown as feed for livestock.

In the north of England and Scotland, the turnip is called *neep*; the word *turnip* itself is an old compound of *neep*. *Neep* often also refers to the larger, yellow rutabaga root vegetable which is also known as the "swede" (from "Swedish turnip").

Description

The most common type of turnip is mostly white-skinned apart from the upper 1–6 centimetres, which protrude above the ground and are purple, red, or greenish wherever sunlight has fallen. This above-ground part develops from stem tissue, but is fused with the root. The interior flesh is entirely white. The entire root is roughly conical, but can be occasionally global, about 5–20 centimetres in diameter, and lacks side roots. The taproot (the normal root below the swollen storage root) is thin and 10 centimetres or more in length; it is trimmed off before marketing. The leaves grow directly from the above-ground shoulder of the root, with little or no visible crown or neck (as found in rutabagas).

Turnip leaves are sometimes eaten as "turnip greens" ("turnip tops" in the UK), and they resemble mustard greens in flavour. Turnip greens are a common side dish in southeastern US cooking, primarily during late fall and winter. Smaller leaves are preferred; however, any bitter taste of larger leaves can be reduced by pouring off the water from initial boiling and replacing it with fresh water. Varieties specifically grown for the leaves resemble mustard greens more than those grown for the roots, with small or no storage roots. Varieties of *B. rapa* that have been developed only for the use of leaves are called Chinese cabbage. Both leaves and root have a pungent flavour similar to raw cabbage or radishes that becomes mild after cooking. Turnip roots weigh up to about one kilogram, although they can be harvested when smaller. Size is partly a function of variety and partly a function of the length of time the turnip has grown. Most very small turnips (also called baby turnips) are speciality varieties. These are only available when freshly harvested and do not keep well. Most baby turnips can be eaten whole, including their leaves. Baby turnips come in yellow-, orange-, and red-fleshed varieties as well as white-fleshed. Their flavour is mild, so they can be eaten raw in salads like radishes and other vegetables.

Nutrition

The turnip's root is high in vitamin C. The green leaves of the turnip top ("turnip greens") are a good source of vitamin A, folate, vitamin C, vitamin K and calcium. Turnip greens are high in lutein (8.5 mg / 100 g). One medium raw turnip (122 g) contains the following nutritional information according to the USDA:

- Calories : 34
- Fat: 0.12
- Carbohydrates: 7.84
- Fibres: 2.2
- Protein: 1.10
- Cholesterol: 0

Like rutabaga, turnip contains bitter cyanoglucosides that release small amounts of cyanide. Sensitivity to the bitterness of these cyanoglucosides is controlled by a paired gene. Subjects who have inherited two copies of the "sensitive" gene find turnips twice as bitter as those who have two "insensitive" genes, and thus may find turnips and other cyanoglucoside-containing foods intolerably bitter.

Origin

There is evidence that the turnip was domesticated before the 15th century BC; it was grown in India at this time for its oil-bearing seeds. The turnip was a well-established crop in Hellenistic and Roman times, which leads to the assumption that it was brought into cultivation earlier. Sappho, a Greek poet from the 7th century BC, calls one of her paramours *Gongýla*, "turnip". Zohary and Hopf note, however, "there are almost no archaeological records available" to help determine its earlier history and domestication. Wild forms of the hot turnip and its relatives the mustards and radishes are found over west Asia and Europe, suggesting their domestication took place somewhere in that area. However Zohary and Hopf conclude, "Suggestions as to the origins of these plants are necessarily based on linguistic considerations."

Cultivation

The 1881 *Household Cyclopedia* gives these instructions for field cultivation of turnips:

The benefits derived from turnip husbandry are of great magnitude; light soils are cultivated with profit and facility; abundance of food is provided for man and beast; the earth is turned to the uses for which it is physically calculated, and by being suitably cleaned with this preparatory crop, a bed is provided for grass seeds, wherein they flourish and prosper with greater vigor than after any other preparation.

Figure: *Turnip (flower)*

The first ploughing is given immediately after harvest, or as soon as the wheat seed is finished, either in length or across the field, as circumstances may seem to require. In this state the ground remains till the oat seed is finished, when a second ploughing is given to it, usually in a contrary direction to the first. It is then repeatedly harrowed, often rolled between the harrowings and every particle of root-weeds carefully picked off with the hand; a third ploughing is then bestowed, and the other operations are repeated. In this stage, if the ground has not been very foul, the seed process.

The next part of the process is the sowing of the seed; this may be performed by drilling machines of different sizes and constructions, through all acting on the same principle. A machine drawn by a horse in a pair of shafts, sows two drills at a time and answers extremely well, where the ground is flat, and the drills properly made up. The weight of the machine ensures a regularity of sowing hardly to be gained by those of a different size and construction. From two to three pounds of seed are sown upon the acre (2 to 3 kg/hectare), though the smallest of these quantities will give many more plants in ordinary seasons than are necessary; but as the seed is not an expensive article the greater part of farmers incline to sow thick, which both provides against the danger of part of the seed perishing, and gives the young plants an advantage at the outset.

Turnips are sown from the beginning to the end of June, but the second and third weeks of the month are, by judicious farmers, accounted the most proper time. Some people have sown as early as May, and with advantage, but these early fields are apt to run to seed before winter, especially if the autumn be favourable to vegetation. As a general rule it may be laid down that the earliest sowings should be on the latest soils; plants on such soils are often long before they make any great progress, and, in the end, may be far behind those in other situations, which were much later sown. The hot turnip plant, indeed, does not thrive rapidly till its roots reach the dung, and the previous nourishment afforded them is often so scanty as to stunt them altogether before they get so far.

The first thing to be done in this process is to run a horse-hoe, called a scraper, along the intervals, keeping at such a distance from the young plants that they shall not be injured; this operation destroys all the annual weeds which have sprung up, and leaves the plants standing in regular stripes or rows. The hand hoeing then commences, by which the turnips are all singled out at a distance of from 8–12 inches, and the redundant ones drawn into the spaces between the rows. The singling out of the young plants is an operation of great importance, for an error committed in this process can hardly be afterward rectified. Boys and girls are always employed as hoers; but a steady and trusty man-servant is usually set over them to see that the work is properly executed.

In eight or ten days, or such a length of time as circumstances may require, a horse-hoe of a different construction from the scraper is used. This, in fact, is generally a small plough, of the same kind with that commonly wrought, but of smaller dimensions. By this implement, the earth is pared away from the sides of the drills, and a sort of new ridge formed in the middle of the former interval. The hand-hoers are again set to work, and every weed and superfluous turnip is cut up; afterward the horse-hoe is employed to separate the earth, which it formerly threw into the furrows, and lay it back to the sides of the drills. On dry lands this is done by the scraper, but where the least tendency to moisture prevails, the small plough is used, in order that the furrows may be perfectly cleaned out. This latter mode, indeed, is very generally practiced.

As a root crop, turnips grow best in cool weather; hot temperatures cause the roots to become woody and bad-tasting. They are typically planted in the spring in cold-weather climates (such as the northern US and Canada) where the growing season is only 3–4 months. In temperate climates (ones with a growing season of 5–6 months), turnips may also be planted in late summer for a second fall crop. In warm-weather climates (7 or more month growing season), they are planted in the fall. 55–60 days is the average time from planting to harvest.

Turnips are a biennial plant, taking two years from germination to reproduction. The root spends the first year growing and storing nutrients, and the second year flowers, produces seeds, and dies. The flowers of the turnip are tall and yellow, with the seeds forming in pea-like pods. In areas with less than 7 month growing seasons, temperatures are too cold for the roots to survive the winter months. In order to produce seeds, it's necessary to pull the turnip and store it overwinter, taking care not to damage the leaves. During the spring, it may be set back in the ground to complete its life cycle.

Human Use

Pliny the Elder considered the turnip one of the most important vegetables of his day, rating it "directly after cereals or at all events after the bean, since its utility surpasses that of any other plant". Pliny praised it as a source of fodder for farm animals, noting that this vegetable is not particular about the type of soil it grows in and that - because it can be left in the ground until the next harvest - it "prevents the effects of famine" for humans (*N.H.* 18.34).

The Macomber turnip (actually a rutabaga) dating from the late 19th century features in one of the very few historic markers for a vegetable, on Main Road in Westport, Massachusetts. In England, around 1700, Turnip Townshend promoted the use of turnips in a four-year crop-rotation system that enabled year-round livestock production. In the south of England and Scotland, the smaller white vegetable is often called *neeps* or *turnips,* while it is the larger rutabagas which are referred to as *swedes.* Turnips or neeps are mashed and eaten with haggis, traditionally on Burns Night.

Turnip lanterns are an old tradition; since inaugural Halloween festivals in Ireland and Scotland, turnips (rutabaga) have been carved out and used as candle lanterns. At Samhain, candle lanterns carved from turnips — *samhnag* — were part of the traditional Celtic festival. Large turnips were hollowed out, carved with faces and placed in windows, used to ward off harmful spirits. At Halloween in Scotland in 1895, masqueraders in disguise carried lanterns made out of scooped out turnips.

In Nordic countries turnips provided the staple crop before their replacement by the potato in the 18th century. The cross between turnip and cabbage, rutabaga, was possibly first produced in Scandinavia.

In Turkey, particularly in the area near Adana, turnips are used to flavour °aljam, a juice made from purple carrots and spices served ice cold. In Middle Eastern countries such as Lebanon, turnips are pickled.

In Japan pickled turnips are also popular and are sometimes stir fried with salt/soysauce. Turnip greens are included in the ritual of the Festival of Seven Herbs, called *suzuna.*

In the Southern United States, stewed turnips are eaten as a root vegetable in the autumn and winter. The greens of the turnip are harvested and eaten all year. Turnip greens may be cooked with a ham hock or piece of fat pork meat, the juice produced in the stewing process prized as pot liquor. Stewed turnip greens are often eaten with vinegar.

In the Tyrolean Alps of Austria, raw shredded turnip-root is served in a chilled remoulade in the absence of other fresh greens as a winter salad.

In Iran, boiled turnip-roots (with salt) are a common household remedy for cough and cold.

In the Punjab and Kashmir regions of India and Pakistan turnips are used in variety of dishes, most notably shab-daig.

The low popularity of turnips (*nabos*) in Brazil, traditionally regarded as distasteful, or at least somewhat disagreeable and unpleasant to the first bite, and that – to the popular knowledge – only feature significantly in the country's Japanese cuisine (as daikon – actually a big and white variety of radish, vegetable there known as *rabanete*), is often a source of humor. Part of this bias reportedly stems from the Middle Ages, where, for the reason of being inexpensive, turnips became [in Iberia and thus in Iberian-descended cultures] associated with the poor, and avoided in the diet of the nobility.

Heraldry

The turnip is an old vegetable charge in heraldry. It was used by Leonhard von Keutschach, prince-archbishop of Salzburg.

Common Pests that Attack Turnips

There aren't many pests that attack turnips, but some insects do find this root crop yummy and when they do, the damage can be pretty significant. Insect pests can do direct and indirect harm to turnips or any other vegetable for that matter. Some insects feed on the fruit and do direct damage. Some do indirect damage by eating the leaves or sucking the plant's sap.

Although in some cases, the harm is not that extensive or directly damaging, the plant still is affected in some ways. They are unable to produce fruit if they lose too much leaves and some can get contaminated by disease transmitted by these pests.

Turnip Pests

Two different types of flea beetle attack turnips – cabbage flea beetle and striped flea beetle. Both eat holes on turnips' first leaves, chew stems and can do extensive damage if not controlled. Aphids, caterpillars, root maggots and grub also feed on turnips. Aphids, or what we commonly know as turnip louse, feed on the underside of the leaves of turnips making them curl and turn yellow. Vegetable Weevils are leaf-feeders but can also damage the stems and roots. You will see them crawling on plant crowns and leaves.

Marked in striking yellow, black or red, Harlequin Bugs eat collards, cabbage and turnip leaves, leaving sunken spots and blotches on the part where they fed.

Remedy and Control

When pests attack your turnips, worry not. There are steps that can be taken to successfully control or eliminate these pests. In fact, some infestation can be avoided altogether even before they begin to happen. Like with flea beetles. These pests can be prevented by simply applying insecticide to the soil when planting. Covering your turnips with row covers can avoid root maggots from harming your crops.

Malathion and carbaryl will help eliminate vegetable weevils and rotating the garden site, cultivation and tillage can help reduce the level of infestation. You may also choose to grow turnip varieties such as Forage Star which has the ability to resist aphids. Harlequin bugs can be eliminated by hand-picking eggs from small forages. Spraying carbaryl, malathion or permethrin will also help.

Diseases

Turnips are a quick growing plant and because of this they do not suffer too much from pests and diseases but there are some that can require attention.

Club root is a major problem and causes wilting and yellowing of the leaves and lesions on the roots. The problem usually occurs in soils with a PH value below 7.0 so test the ground and keep the PH level up and will occur where the turnips have been repeatedly planted in the same area. Once in the ground the disease will stay for 7 years so avoid planting in the same area again. Crop rotation is essential on at least a 4 year cycle.

Root Maggots are small bugs that will eat through the tubers and ruin them. They become prevalent at certain times of the year. The problem can be either cured or alleviated by dowsing them in a suitable insecticide. This should be timed to coincide with the time the eggs are laid. Crop rotation helps remove the problem.

Aphids can get under the leaves but a good ladybird population will sort it, if not wash the leaves with soapy water.

Wireworm can sometimes be found and the cure is the same as the root beetle with a suitable insecticide. Crop rotation helps remove the problem. Fungal problems can occur and these can be washed of with soapy water or a suitable fungicide.

Generally storage needs to be cool and dry to stop any rotting or wilting of the tubers but they can be left in the ground over the winter and used when required.

The soil should ideally be warm when planting to get the best results (60 degrees C) and neutral, not acidic and crop rotation is essential.

Mushroom

Edible mushrooms are the fleshy and edible fruit bodies of several species of macrofungi (fungi which bear fruiting structures that are large enough to be seen with the naked eye). They can appear either below ground (hypogeous) or above ground (epigeous) where they may be picked by hand. Edibility may be defined by criteria that include absence of poisonous effects on humans and desirable taste and aroma.

Edible mushrooms are consumed by humans as comestibles for their nutritional value and they are occasionally consumed for their supposed medicinal value. Mushrooms consumed by those practicing folk medicine are known as medicinal mushrooms. While hallucinogenic mushrooms (e.g. Psilocybin mushrooms) are occasionally consumed for recreational or religious purposes, they can produce severe nausea and disorientation, and are therefore not commonly considered edible mushrooms.

Edible mushrooms include many fungal species that are either harvested wild or cultivated. Easily cultivatable and common wild mushrooms are often available in markets, and those that are more difficult to obtain (such as the prized truffle and matsutake) may be collected on a smaller scale by private gatherers. Some preparations may render certain poisonous mushrooms fit for consumption.

Before assuming that any wild mushroom is edible, it should be identified. Proper identification of a species is the only safe way to ensure edibility. Some mushrooms that are edible for most people can cause allergic reactions in some individuals, and old or improperly stored specimens can cause food poisoning. Deadly poisonous mushrooms that are frequently confused with edible mushrooms and responsible for many fatal poisonings include several species of the *Amanita* genus, in particular, *Amanita phalloides*, the *death cap*. Mushrooms growing in polluted locations can accumulate pollutants such as heavy metals.

History of Mushroom Use

Mycophagy the act of consuming mushrooms, dates back to ancient times. Edible mushroom species have been found in association with 13,000 year old ruins in Chile, but the first reliable evidence of mushroom consumption dates to several hundred years BC in China. The Chinese value mushrooms for medicinal properties as well as for food. Ancient Romans and Greeks, particularly the upper classes, used mushrooms for culinary purposes. Food tasters were employed by Roman Emperors to ensure that mushrooms were safe to eat. Mushrooms are also easily preserved, and historically have provided additional nutrition over winter. Many cultures around the world have either used or continue to use psilocybin mushrooms for spiritual purposes as well as medicinal mushrooms in folk medicine.

Current Culinary Use

A fraction of the many fungi consumed by humans are currently cultivated and sold commercially. Commercial cultivation is important ecologically, as there have been concerns of depletion of larger fungi such as chanterelles in Europe, possibly because the group has grown so popular yet remains a challenge to cultivate.

Commercially Cultivated

Mushroom cultivation has a long history, with over twenty species commercially cultivated. Mushrooms are cultivated in at least 60 countries with China, the United States, Netherlands, France and Poland being the top five producers in 2000.

Commercially Harvested Wild Edibles

Figure: *Chanterelles in the wild*

Some species are difficult to cultivate; others (particularly mycorrhizal species) have not yet been successfully cultivated. Some of these species are harvested from the wild, and can be found in markets. When in season they can be purchased fresh, and many species are sold dried as well. The following species are commonly harvested from the wild:

- *Boletus edulis* or edible Boletus, native to Europe, known in Italian as Fungo Porcino (plural 'porcini') (Pig mushroom), in German as Steinpilz (Stone mushroom), in Russian as "white mushroom", in Albanian as (Wolf mushroom) and in French the *cèpe*. It also known as the king bolete, and is renowned for its delicious flavour. It is sought after worldwide, and can be found in a variety of culinary dishes.

- *Cantharellus cibarius* (The chanterelle), The yellow chanterelle is one of the best and most easily recognisable mushrooms, and can be found in Asia, Europe, North America and Australia. There are poisonous mushrooms which resemble it, though these can be confidently distinguished if one is familiar with the chanterelle's identifying features.
- *Cantharellus tubaeformis*, the tube chanterelle or yellow-leg
- *Clitocybe nuda* - Blewit (or Blewitt)
- *Cortinarius caperatus* the Gypsy mushroom (recently moved from genus *Rozites*)
- *Craterellus cornucopioides* - Trompette de la Mort or Horn of Plenty
- *Grifola frondosa*, known in Japan as *maitake* (also "hen of the woods" or "sheep's head"); a large, hearty mushroom commonly found on or near stumps and bases of oak trees, and believed to have *Macrolepiota procera* properties.
- *Gyromitra esculenta* this "False morel" is prized by the Finns. This mushroom is deadly poisonous if eaten raw, but highly regarded when parboiled.
- *Hericium erinaceus*, a tooth fungus; also called "lion's mane mushroom."
- *Hydnum repandum* Sweet tooth fungus, hedgehog mushroom, urchin of the woods
- *Lactarius deliciosus* Saffron milk cap - Consumed around the world and prized in Russia
- *Morchella* species, (morel family), morels belong to the ascomycete grouping of fungi. They are usually found in open scrub, woodland or open ground in late spring. When collecting this fungus, care must be taken to distinguish it from the poisonous false morels, including *Gyromitra esculenta.*
    - o *Morchella conica var. deliciosa*
    - o *Morchella esculenta var. rotunda*
- *Tricholoma matsutake* the Matsutake, a mushroom highly prized in Japanese cuisine.
- *Tuber* species, (the truffle), Truffles have long eluded the modern techniques of domestication known as *trufficulture*. Although

the field of trufficulture has greatly expanded since its inception in 1808, several species still remain uncultivated. Domesticated truffles include

- o *Tuber borchii*
- o *Tuber brumale*
- o *Tuber indicum* - Chinese black truffle
- o *Tuber macrosporum* - White truffle
- o *Tuber mesentericum* - The Bagnoli truffle
- o *Tuber uncinatum* - Black summer truffle

Other Edible Wild Species

Many wild species are consumed around the world. The species which can be identified "in the field" (without use of special chemistry or a microscope) and therefore safely eaten vary widely from country to country, even from region to region. This list is a sampling of lesser-known species that are reportedly edible.

- *Amanita caesarea* (Caesar's Mushroom)
- *Armillaria mellea*
- *Boletus badius*
- *Chroogomphus rutilus* (pine-spikes or spike-caps)
- *Calvatia gigantea* (Giant Puffball)
- *Clavariaceae* species (coral fungus family)
- *Clavulinaceae* species (coral fungus family)
- *Coprinus comatus*, the Shaggy mane. Must be cooked as soon as possible after harvesting or the caps will first turn dark and unappetizing, then deliquesce and turn to ink. Not found in markets for this reason.
- *Corn smut*
- *Cortinarius variicolour*
- *Fistulina hepatica* (beefsteak polypore or the ox tongue)
- *Hygrophorus chrysodon*
- *Lactarius salmonicolour*
- *Lactarius subdulcis* (mild milkcap)
- *Lactarius volemus*
- *Laetiporus sulphureus* (Sulphur shelf). Also known by names such as the "chicken mushroom", "chicken fungus", sulphur

shelf is a distinct bracket fungus popular among mushroom hunters.

- *Leccinum aurantiacum* (Red-capped scaber stalk)
- *Leccinum scabrum* (Birch bolete)
- *Lepiota procera*
- *Macrolepiota procera* Parasol Mushroom - Globally, it is widespread in temperate regions
- *Polyporus squamosus* (Dryad's saddle and Pheasant's back mushroom)
- *Polyporus mylittae*
- *Ramariaceae* species (coral fungus family)
- *Rhizopogon luteolus*
- *Russula*, some members of this genus are edible.
- *Sparassis crispa*. Also known as "cauliflower mushroom".
- *Suillus bovinus*
- *Suillus granulatus*
- *Suillus luteus*
- *Suillus tomentosus*
- *Tricholoma terreum*

Conditionally Edible Species

There are a number of fungi that are considered choice by some and toxic by others. In some cases, proper preparation can remove some or all of the toxins.

- *Amanita muscaria* is edible if parboiled to leach out toxins. Fresh mushrooms cause vomiting, twitching, drowsiness, and hallucinations due to the presence of muscimol. Although present in *A. muscaria*, ibotenic acid is not in high enough concentration to produce any physical or psychological effects unless massive amounts are ingested.
- *Coprinopsis atramentaria* is edible without special preparation. However, consumption with alcohol is toxic due to the presence of coprine. Some other *Coprinus* spp. share this property.
- *Gyromitra esculenta* is eaten by some after it has been parboiled; however, mycologists do not recommend it. Raw *Gyromitra* are toxic due to the presence of gyromitrin, and it is not known if all of the toxin can be removed by parboiling.

- *Lactarius spp.* - Apart from *Lactarius deliciosus* which is universally considered edible, other *Lactarius spp.* that are considered toxic elsewhere in the world are eaten in Russia after pickling or parboiling.
- *Verpa bohemica* - Considered choice by some, it even can be found for sale as a "morel", but cases of toxicity have been reported. Verpas contain toxins similar to gyromitrin and similar precautions apply.

Use in Traditional Medicine

Medicinal mushrooms are mushrooms or extracts from mushrooms that are thought to be treatments for diseases, yet remain unconfirmed in mainstream science and medicine, and so are not approved as drugs or medical treatments. Such use of mushrooms therefore falls into the domain of traditional medicine.

Figure: *Reishi, a well-known mushroom*

Preliminary research has shown some medicinal mushroom isolates to have cardiovascular, anticancer, antiviral, antibacterial, antiparasitic, anti-inflammatory, and antidiabetic properties. Currently, several extracts (polysaccharide-K, polysaccharide peptide, lentinan) have widespread use in Japan, Korea and China, as potential adjuvants to radiation treatments and chemotherapy.

The concept of a medicinal mushroom has a history spanning millennia in parts of Asia, mainly as traditional Chinese medicine. Only a few mushroom extracts have been tested for potential efficacy. Preliminary research results for most other extracts are on isolated cell lines, animal research with rodents, or early-stage clinical human trials as pilot studies.

Pests and Diseases of Mushrooms

Invertebrate Pests

Arthropods: A variety of small fly and midge species are pests of mushrooms. The larvae feed on the fungal mycelium in the compost, but may also tunnel into the fruiting bodies.

A range of mite species may affect the mushroom crop. Some directly damage the fruiting bodies, some may attack the mycelium and some mites are predatory on other mites, fly eggs, nematodes or bacteria. Mite damage on the fruiting bodies often shows up as small cavities in the stem and cap similar in appearance to bacterial pit disease. Mycelium-eating mites can cause high yield losses. Mites are very small and easily transported on clothing and tools.

Springtails are commonly associated with compost and can damage the crop if present in high enough numbers. Slaters and millipedes may also cause damage to the fruiting bodies.

Nematodes

Nematodes will cause a loss in yield and brown, watery mushrooms and, in extreme cases, a soggy, smelly compost. Peat is a common source for nematodes and should be treated before use.

Bacterial Diseases

Pseudomonas (bacterial spot, bacterial or pit or brown blotch) cause yellow to brown blotches on the cap, which may exude sticky residues. Early symptoms are similar to verticilium.

Management

Maintaining high levels of hygiene will assist any pest management program by reducing the number of problems that are likely to occur. If an outbreak does occur, ensure that you correctly identify the pest or disease before taking any action.

Taro

Taro is a common name for the corms and tubers of several plants in the Araceae family. Of these, *Colocasia esculenta* is the most widely cultivated and the subject of this article. More specifically, this article describes the "dasheen" form of taro; another variety of taro is known as *eddoe*. Taro is native to South India and Southeast Asia. It is a perennial, tropical plant primarily grown as a root vegetable for its edible starchy corm, and as a leaf vegetable. It is a food staple in African, Oceanic and South Indian cultures and is believed to have been one of the earliest cultivated plants. *Colocasia* is thought to have originated in the Indo-Malayan region, perhaps in eastern India and Bangladesh, and spread eastward into Southeast Asia, eastern Asia, and the Pacific islands; westward to Egypt and the eastern Mediterranean; and then southward and westward from there into East Africa and West Africa, whence it spread to the Caribbean and Americas. It is known by many local names and often referred to as "elephant ears" when grown as an ornamental plant.

Figure: *Taro*

Cultivation

Taro can be grown in paddy fields where water is abundant or in upland situations where water is supplied by rainfall or supplemental irrigation. Taro is one of the few crops (along with rice and lotus) that can be grown under flooded conditions. This is due to air spaces in the petiole, which permit underwater gaseous exchange with the atmosphere. For a maximum dissolved oxygen supply, the water should be cool and flowing. Warm, stagnant water causes basal rotting. For maximum yields, the water level should be controlled so that the base of the plant is always under water.

Flooded cultivation has some advantages over dry-land cultivation: higher yields (about double), out-of-season production (which may result in higher prices), and weed control (which flooding facilitates). On the other hand, in flooded production systems taro requires a longer maturation period, investment in infrastructure, and higher operational costs, and monoculture is likely.

Like most root crops, taro and eddoes do well in deep, moist or even swampy soils where the annual rainfall exceeds 2,500 mm. Eddoes are more resistant to drought and cold. The crop attains maturity within six to twelve months after planting in dry-land cultivation and after twelve to fifteen months in wetland cultivation. The crop is harvested when the plant height decreases and the leaves turn yellow. These signals are usually less distinct in flooded taro cultivation.

Harvesting is usually done by hand tools, even in mechanized production systems. First, the soil around the corm is loosened, and then, the corm is pulled up by grabbing the base of the petioles. The global average yield is 6.2 tones/hectare but varies according to the region. In Asia, average yields reach 12.6 tones/hectare.

Toxicity

The plant is inedible when raw and considered toxic due to the presence of calcium oxalate crystals, typically as raphides. The toxin is minimized by cooking, especially with a pinch of baking soda. It can also be reduced by steeping taro roots in cold water overnight. Calcium oxalate is highly insoluble and contributes to kidney stones. It has been recommended to consume milk or other calcium-rich foods together with taro.

Culinary Use

The corms, which have a light purple colour due to phenolic pigments, are roasted, baked or boiled, and the natural sugars give a sweet nutty flavour. The starch is easily digestible, and since the grains are fine and small it is often used for baby food. The leaves are a good source of vitamins A and C and contain more protein than the corms.

Azores

In the Azores taro is known as *inhâme* or *inhame-coco* and is commonly steamed with potatoes, vegetables and meats or fish. It is also consumed as a dessert after being fried in vegetable oil or lard, after steaming and peeling, and then sprinkled with sugar. Taro grows in the fertile land of the Azores, as well as in creeks that are fed by mineral springs. Through immigration, the *inhâme* is found in the Azorean diaspora.

Bangladesh

In Bangladesh taro is a very popular vegetable known as *mukhi* or *mukhi kochu* . It is usually cooked with small prawns or the ilish fish into a curry, but some dishes are cooked with dried fish. Its green leaves, *kochu pata* , and stem, *kochu* , are also eaten as a favourite dish and usually ground to a paste or finely chopped to make *saag* - but it must be boiled well beforehand. Taro stolons or stems, *kochur loti* , are also favoured by Bangladeshis and cooked with shrimps, dried fish or the head of the ilish fish. Taro is available, either fresh or frozen, in the UK and US in most Asian stores and supermarkets specialising in Bangladeshi or South Asian food. Also, another variety called *maan kochu* is consumed and is a rich source of vitamins and nutrients.

Brazil

In Portuguese-speaking countries, *inhame* (pronounced or, literally "yam") and *cará* are the common names for various plants with edible parts of the genera *Alocasia, Colocasia* (family Araceae) and *Dioscorea* (family Dioscoreaceae), and its respective starchy edible parts, generally tubers, with the exception of the air potato *Dioscorea bulbifera,* called *cará-moela* (pronounced, literally, "gizzard yam"), in Brazil and never deemed to be an *inhame.* Definitions of what constitutes an *inhame* and a *cará* vary regionaly, but the common understanding in Brazil is that *carás* are potato-like in shape, while *inhames* are more oblong.

In the "broad" lower class Brazilian Portuguese (this sociocultural variation is sometimes considered comparable to that of Australian English) of the hotter and drier Northeastern region, both *inhames* and *carás* are called *batata* (literally, "potato"). For differentiation, potatoes are called *batata-inglesa* (literally, "English potato"), a name used in other regions and sociolects to differentiate it from the *batata-doce,* "sweet potato", ironic names since both were first cultivated by the indigenous peoples of South America, their native continent, and only later introduced in Europe by the colonizers. Taros are often prepared like potatoes, eaten boiled, stewed or mashed, generally with salt and sometimes garlic as condiment, as part of a meal (most often lunch or dinner).

Chinese Cuisine

China: Taro is commonly used as a main course (steamed taro with or without white sugar, as a substitute for other cereals), in Chinese cuisine in a variety of styles and provinces steamed, boiled or stir fried as a main dish and as a flavour enhancing ingredient. In Northern China it is often boiled or steamed then peeled and eaten with or without sugar much like a potato. It is commonly braised with pork or beef. It is used in the dim sum cuisine of southern China to make a small plated dish called taro dumpling as well as a pan-fried dish called taro cake. It is also shredded into long strips which are woven together to form a seafood birdsnest. Taro cake is a delicacy traditionally eaten during Chinese New Year celebrations. In desserts it is used in *tong sui*, bubble tea, as a flavouring in ice cream and other desserts in China (e.g., Sweet Taro Pie). McDonald's sells taro-flavoured pies in China.

Taiwan

In Taiwan, where taro is called *ôX-á* (Taiwanese Hokkien: ‹,ÔN), supermarket varieties range from about the size and shape of a brussels sprout to longer, larger varieties the size of a person's fist, or even bigger. Taro chips are often used as a potato chip-like snack. Compared to potato chips, taro chips are harder and have a more nutty flavour. Other popular traditional Taiwanese taro snacks are deep-fried taro balls. Taro is prominently featured in Taiwanese cold desserts.

Cook Islands

There are many taro plantations on the Cook Islands as the soil there is perfect for them. The root is eaten boiled, as is standard across many Polynesian islands. The leaves are also eaten as a delicacy (known locally as Rukau or "greens"), cooked with coconut milk, onion and meat or fish.

Costa Rica

In Costa Rica, taro is eaten in soups, as a replacement for potatoes, and as chips. It is known locally as "Tiquisque".

East Africa

In Kenya, Uganda and Tanzania, taro is commonly known as *Arrow root*, or *Nduma* in some local Bantu languages. It is usually boiled and eaten with tea or other beverages, or as the main starch of a meal.

Egypt

In Egypt, Taro is Known as Kolkas: The corms are larger than what would be found in North American supermarkets. After being peeled completely, it is cooked in one of two ways. It is cut into small cubes and cooked in broth with fresh coriander and swiss chard and served as an accompaniment to meat stew, or it may be sliced and cooked with minced meat and tomato sauce.

Those who have gone through compulsory army conscription remember the much reviled *black dish* which is a combination of eggplants and unpeeled taro halves.

Europe

Taro was consumed by the early Romans in much the same way the potato is today. They called this root vegetable *colocasia*. The Roman cookbook Apicius mentions several methods for preparing taro, including boiling, preparaing with sauces, and cooking with meat or fowl. After the fall of the Roman Empire the use of taro dwindled in Europe. This was largely due to the decline of trade and commerce with Egypt, previously controlled by Rome. It is still important to note the Taro because when the Spanish and Portuguese sailed to the new world, they brought taro along with them. Taro has remained popular in the Canary Islands. Recently there has been renewed interest in exotic foods and consumption is increasing.

Cyprus

In Cyprus, taro has been in use since the time of the Roman Empire. Today it is known as *kolokasi*, which is similar to the name the Romans used: *colocasia*. It is usually sauteed with celery and onion with pork or chicken, in a tomato sauce - a vegetarian version is also available. "Baby" taro is called "poulles" on the island, and after being sauteed the vessel is decaramelised with dry red wine and coriander seeds, then served with freshly squeezed lemon.

Greece

In Greece, taro grows on Ikaria island. Ikarians credit taro for saving them from famine during World War II. They boil it until tender and serve it as a salad.

Spain

Taro root is called *ñame* in Spanish and is largely cultivated in the Autonomous Community of the Canary Islands (Canary Islands, Spain).

Fiji

Taro (*dalo* in Fijian) has been a staple of the Fijian diet for centuries, and its cultural importance is celebrated on Taro Day. Its growth as an export crop can be said to have begun in 1993 when the taro leaf blight (*Phytophthora colocasiae*) decimated the taro industry in neighbouring Samoa. Fiji filled the void and was soon supplying taro internationally. Almost 80% of Fiji's exported taro comes from the island of Taveuni where the taro beetle species (*Papuana uninodis*) is absent. The Fijian taro industry on the main islands of Viti Levu and Vanua Levu faces constant damage from the beetles. The Fiji

Ministry of Agriculture and the Land Resources Division of the Secretariat of the Pacific Community (SPC) are researching pest control and instigating quarantine restrictions to prevent spread of the pest. Taveuni now exports pest damage-free crops.

India

In India, taro or eddoe is a common dish served in many ways.

In Manipur, the north-eastern state of India, taro is known as "pan". It is widely available and is eaten in many forms, either baked, boiled, and cooked into a curry with Hilsa fish or with fermented soyabeans called "Hawai-zaar". The leaves are also used in a special traditional dish called "utti", cooked with peas.

In Himachal Pradesh, a northern state in India, taro is known as *ghandyali* in Mandi district. The dish called *patrodu* is made from the leaves of the *ghandyali*. Also in the capital Shimla, a pancake-style dish, called *patra* or *patid*, is made using gram flour.

A tall-growing variety of taro is extensively used on the western coast of India to make *patrode*, *patrade*, or *patrada*, literally a "leaf-pancake". In Dakshin Kannada district in the state of Karnataka, it is used as a morning breakfast dish, either made like fritters, or steamed. In the state of Maharashtra, the leaves, called *alu che paana*, are de-veined, rolled with a paste of gram flour, tamarind paste, red chilli powder, turmeric, coriander, asoefotida, and salt, and then steamed. These can be eaten whole or cut into pieces, or shallow fried and eaten as a snack known as *alu chi wadi*. *Alu chya panan chi patal bhaji*, a lentil and colocasia leaves curry, is also popular. In the Indian state of Gujarat, the leaves of the plant are used to make patra, a dish with tamarind and other spices. Sindhis call it *kachaloo*; they fry it, mash it, and re-fry it to make a dish called *took* which complements Sindhi curry.

In Kerala, a state in southern India, taro corms are known as *chembu-kizhangu*. Taro is used as a staple food, as a side dish, or as an ingredient in various side dishes like sambar. As a staple food, it is steamed and eaten with a spicy chutney of green chillies, tamarind, and shallots. The leaves and stems of certain varieties of taro are used as a vegetable in Kerala.

In other Indian states, Tamil Nadu and Andhra Pradesh, taro corms are known as *sivapan-kizhangu* (seppankilangu or cheppankilangu), *chamagadda*, or in coastal Andhra districts as *chaama dumpa* in Telugu, and it can be cooked in many ways, such as deep fried in oil for a side item with rice, or cooked in a tangy tamarind sauce with spices, onion, and tomato.

In the East Indian state of West Bengal, taro roots are thinly sliced and fried to make chips called *kochu bhaja*. The stem is used to cook a very tasty *saag*, often eaten as a starter with hot rice. The roots are also made into a paste with spices and eaten with rice. The most popular dish is a spicy curry made with prawn and taro roots.

In the eastern Indian state of Odisha, taro root is known as *saru*. Dishes made of taro include *saru besara* (taro in mustard and garlic paste). It is also an indispensable ingredient in preparing the heart of Oriya cuisine, the *dalma*, where vegetables are cooked with dal. Sliced taro roots, deep fried in oil and mixed with red chili powder and salt, are known as *saru chips*.

In the Indian state of Uttarakhand and neighbouring Nepal, taro is considered a healthy food cooked in a variety of ways. The kumaon region of uttarakhand calls it *gaderi*. The most common style is boiling it in salty water in iron cooking pots until it becomes like porridge. Another style is to steam the young leaves, called *gava*, then to sundry and store them for later use. For another use, the taro leaves and stems are used raw as an ingredient for pickles. The leaves and stems are mixed with black lentils and then dried as small balls called *badi*. The stems may also be sun-dried and stored for later use. On one special day, women worship *saptarshi* ("seven sages") and eat only rice with taro leaf vegetable.

Japan

In Japan, it is called *satoimo*, the "child" and "grandchild" corms (cormels, cormlets) which bud from the parent *satoimo*, are called *koimo* and *magoimo*, respectively, or more generally *imonoko*. *Satoimo* has been propagated in Southeast Asia since the late Jômon period. It was a regional staple food before rice became predominant. The tuber, *satoimo*, is often prepared through simmering in fish stock (dashi) and soy sauce. The stalk, *zuiki*, can also be prepared a number of ways, depending on the variety.

Lebanon

In Lebanon, taro is known as "kilkass" and is mainly grown along the Mediterranean coast. The leaves and stems are not consumed in Lebanon and the variety grown produces round to slightly oblong tubers that vary in size from a tennis ball to a small cantaloupe. Kilkass is a very popular winter dish in Lebanon and is prepared in two ways: "kilkass with lentils" is a stew flavoured with crushed garlic and lemon juice and "kilkass in tahini".

Another common method of preparing taro is to boil, peel then slice it into 1 cm thick slices, before frying and marinating in edible "red" sumac. In northern Lebanon, it is known as a potato with the name *Borshoushi* (El-Orse Borshushi). It is also prepared as part of a lentil soup with crushed garlic and lemon juice. The smaller variety is more popular due to its tenderness.

Maldives

Ala was widely grown in southern atolls Addu, Fuvahmulah, Huvadhu, and Haddhunmathi and considered a staple food even after rice was introduced. Ala and olhu ala are still widely eaten all over Maldives, cooked or steamed with salt to taste and eaten with grated coconut along with chili paste and fish soup. It is also prepared as a curry. The roots are sliced and fried to make chips and are also used to prepare varieties of sweets.

Nepal

Taro is grown in the hilly regions of Nepal. The root (corm) of taro is known as *pindalu* and petioles with leaves are known as *karkalo* and also as *Gava*. Almost all parts are eaten in different dishes. Boiled corm of Taro is commonly served with salt, spices and chillies. Taro is a popular dish in hilly region. Chopped leaves and petioles are mixed with Urad bean flour to make dried balls called *maseura*. Large taro leaves are used as an alternative to an umbrella when unexpected rain occurs. Popular attachment to taro since ancient times is reflected in popular culture, such as in songs and textbooks. *Jivan hamro karkala ko pani jastai ho* means, "Our life is as vulnerable as water stuck in the leaf of taro".

Philippines

Taro is called *gabi* in the Philippines and is widely available throughout the archipelago. The leaves, stems, and corms are consumed and form part of the local cuisine. A popular recipe for taro is *laing* which is heavily inspired by a dish that originated in the Bicol region. The dish's main ingredients are taro leaves (at times including stems) cooked in coconut milk, and salted with fermented shrimp or fish *bagoong*. It is sometime heavily spiced with red hot chilies called *siling labuyo*. Another dish in which taro is commonly used is the Philippine national stew, called *sinigang*. This stew is made with pork and beef, shrimp, or fish, a souring agent (tamarind leaves, kamias, or vinegar, etc.) with the addition of peeled and diced corms as thickener. The corm is also prepared as a basic ingredient for *ginataan*, a coconut milk and taro desert.

Pakistan

In Pakistan, taro or eddoe or arvi is a very common dish served with or without gravy; a popular dish is *arvi gosht*, which includes beef, lamb or mutton. The leaves are rolled along with gram flour batter and then fried or steamed to make a dish called Pakora, which is finished by tempering with red chillies and carrom (ajwain) seeds.

Papua New Guinea

Among the Urapmin people of Papua New Guinea, taro (known in Urap as *ima*) is the main source of sustenance along with the sweet potato (Urap: *wan*). In fact, the word for "food" in Urap is a compound of these two words.

Polynesia

Considered the staple starch of traditional Polynesian cuisine, taro is both a common and prestigious food item that was first introduced to the Polynesian islands by prehistoric seafarers of Southeast Asian derivation. The tuber itself is prepared in various ways, including baking, steaming in earth ovens (*umu* or *imu*), boiling, and frying. The famous Hawaiian staple poi is made by mashing steamed taro roots with water. Taro also features in traditional desserts such as Samoan "fa'ausi", which consists of grated, cooked taro mixed with coconut milk and brown sugar. The leaves of the taro plant also feature prominently in Polynesian cooking, especially as edible wrappings for dishes such as Hawaiian laulau, Fijian and Samoan "palusami" (wrapped around onions and coconut milk), and Tongan "lupulu" (wrapped corned beef). Ceremonial presentations on occasion of chiefly rites or communal events (weddings, funerals, etc.) traditionally included ritual presentation of raw and cooked taro roots/plants. The Hawaiian laulau traditionally contains pork, fish, and lu'au (cooked taro leaf). The wrapping is inedible ti leaves (Hawaiian: lau ki). Cooked taro leaf has the consistency of cooked spinach and is therefore unsuitable for use as a wrapping.

Samoa

In Samoa the taro root and parcels of coconut milk wrapped in the taro leaves are cooked, along with other food, in an umu. The parcels are called palusami or lu'au. The root is used to scrape up pieces of lu'au so as to eat them together. The resulting mouthful is smoky, sweet, savoury and has a unique taste from the taro leaves. The texture is also remarkable: starchy, creamy, and luscious.

South Korea

In South Korea, taro is called *toran* (Korean: "egg from earth"), and the corm is stewed and the leaf stem is stir-fried. Taro roots can be used for medicinal purposes, particularly for treating insect bites. It is made into the Korean traditional soup *toranguk*. Taro stems are often used as an ingredient in yukgaejang.

Sri Lanka

Many varieties are recorded in Sri Lanka, several being edible, others being toxic to humans and therefore not cultivated. Edible varieties (*kiri ala, kolakana ala, gahala, sevel ala*) are cultivated for their corms and leaves. Sri Lankans eat corms after boiling them or making them into a curry with coconut milk. The leaves of only one variety (*kolakana ala*) are eaten.

Suriname

In Suriname, the taro root is called *aroei* by the native Indians and is commonly known as "Chinese tayer". The variety known as "eddoe" is also called Chinese tayer. It is a popular cultivar among the marroon population in the interior, also because it is not adversely affected by high water levels. The "dasheen" variety, commonly planted in swamps, is rare, although appreciated for its taste. The closely related *Xanthosoma* species is the base for the popular Surinamese dish, pom.

Thailand

In Thai cuisine, taro Thai: is used in a variety of ways depending on the region. Boiled taro is readily available in the market packaged in small cellophane bags, already peeled and diced, and eaten as a snack. Pieces of boiled taro with coconut milk are a traditional Thai dessert. Raw taro is also often sliced and deep fried and sold in bags as chips..

Trinidad & Tobago

Callaloo (sometimes calaloo or kallaloo) is a popular Caribbean dish originated from West Africa served in different variants across the Caribbean. The main ingredient is a leaf vegetable, traditionally either amaranth (known by many local names, including callaloo or bhaaji), taro or Xanthosoma. Both are known by many names, including callaloo, coco, tannia, bhaaji, or dasheen bush. Because the leaf vegetable used in some regions may be locally called "callaloo" or "callaloo bush", some confusion can arise among the different vegetables and with the dish itself. Outside of the Caribbean, water spinach is occasionally used. Trinidadians primarily use taro/dasheen bush for callaloo, while Jamaicans and Guyanese use the name callaloo to refer to amaranth, and use it in a plethora of dishes and also a drink ("callaloo juice").

It should be understood that the "callaloo" made in Jamaica is different from the "callaloo" made in Trinidad and Tobago in terms of main ingredient (the leaf used) and other ingredients included (for example, Jamaicans tend to use only callaloo leaf, salt, onions, and scallions, and simply steam the vegetable, while Trinidadians use okra and coconut milk to make an entirely different dish with a different taste and consistency). Callaloo is the one of National Dishes of Trinidad and Tobago.

Turkey

Taro is grown in the south coast of Turkey, especially in Mersin and Antalya. It is boiled in a tomato sauce or cooked with meat, beans and chickpeas.

United States: Mainland United States

In American Chinatowns, people often use taro in Chinese cuisine, though it is not as popular as in Asian and Pacific nations. Since the late 20th century, taro chips have been available in many supermarkets and natural food stores. In the 1920s, dasheen, as it was known, was highly touted by the Secretary of the Florida Department of Agriculture as a valuable crop for growth in muck fields. Fellsmere, Florida, near the east coast, was a farming area deemed perfect for growing dasheen. It was used in place of potatoes and dried to make flour. Dasheen flour was said to make excellent pancakes when mixed with wheat flour.

Hawaii

In Hawaii, taro, or kalo in the Hawaiian language, is a traditional form of food sustenance and nutrition, known from ancient Hawaiian culture. The contemporary Hawaiian diet consists of many tuberous plants, particularly sweet potato and taro. Some of the uses for taro include poi, table taro, taro chips, and luau leaf. In Hawaii, taro is farmed under either dryland or wetland conditions. Taro farming in the Hawaiian islands is especially challenging because of difficulties in accessing fresh water. Taro is usually grown in pondfields known as *lo»i* in Hawaiian. Cool, flowing water yields the best crop. Typical dryland or upland varieties (varieties grown in watered but not flooded fields) in Hawaii are *lehua maoli* and *bun long*, the latter widely known as Chinese taro. *Bun long* is used for making taro chips. *Dasheen* (also called "eddo") is another "dryland" variety of *C. esculenta* grown for its edible corms or sometimes just as an ornamental plant.

The Hawaii Agricultural Statistics Service puts the 10-year median production of taro in the Hawaiian Islands at about 6.1 million pounds (2,800 t; Viotti, 2004). However, 2003 taro production in Hawaii was only 5 million pounds (2,300 t), an all-time low (record keeping started in 1946). The previous low, reached in 1997, was 5.5 million pounds (2,500 t). Despite generally growing demand, production was even lower in 2005: only 4 million pounds, with *kalo* for processing into *poi* accounting for 97.5%. Urbanization has driven down harvests from a high of 14.1 million pounds (6,400 t) in 1948, but more recently the decline has resulted from pests and diseases. A non-native apple snail (*Pomacea canaliculata*) is a major culprit in the current crop decline. Also, a plant rot disease traced to a newly identified species of the fungal genus *Phytophthora* now plagues crops throughout the state. Although pesticides could control both pests to some extent, pesticide use in the pondfields is barred because of the clear opportunity for chemicals to quickly migrate into streams and then into the ocean.

Important aspects of Hawaiian culture revolves around taro cultivation and consumption. For example, the newer name for a traditional Hawaiian feast, luau, comes from the taro. Young taro tops baked with coconut milk and chicken or octopus arms are frequently served at luaus. Also, one cannot fight when a bowl of poi is open. By ancient Hawaiian custom, it is considered disrespectful to fight in front of an elder. One should not raise the voice, speak angrily, or make rude comments or gestures. An open poi bowl is connected to this concept because Haloa (Taro) is the name of the first-born son of the parents who begat the human race. The ancient Hawaiians identified so strongly with taro that the Hawaiian term for family, 'ohana, is derived from the word 'oha, the shoot or sucker which grows from the taro corm. As young shoots grow from the corm, so people grow from their family.

Venezuela

In Venezuela, taro is called *ocumo chino* or "chino" and used in soups and sancochos. Soups contain large chunks of several kinds of tubers, including *ocumo chino*, especially in the eastern part of the country, where West Indian influence is present. It is also used to accompany meats in "parrillas" (barbecue) or fried cured fish where yuca is not available.

"Ocumo" is an indigenous name, and "chino" means Chinese, since people tend to give the "chinese" adjective to any produce that is considered "exotic". "Ocumo" without the Chinese denomination is a tuber from the same family, but without taros inside purplish colour. Ocumo is the Venezuelan name for malanga, so "ocumo chino" means "chinese malanga". Taro is always prepared boiled. No porridge form is known in the local cuisine.

Vietnam

In Vietnam, where taro is called *Khoai môn* or *Khoai sÍ*, it is used as a filling in spring rolls, cakes, puddings, smoothies, soups and other desserts. Taro is used in the T¿t dessert *chè khoai môn* which is sticky rice pudding with taro roots. The stems are also used in soups such as canh chua.

West Africa

Taro is consumed as a staple crop in West Africa, particularly in Ghana, Nigeria and Cameroon. It is called cocoyam in Nigeria, Ghana and Anglophone Cameroon, and macabo in Francophone Cameroon. Cocoyam is often boiled, fried, or roasted and eaten with a sauce. In Ghana, it substitutes plantain in making Fufu when plantains are out of season. It is also cut into small pieces to make a soupy baby food and appetizer called *mpotompoto*. It also common in Ghana to find Cocoyam chips (deep-fried slices, about 1 mm thick).

Cocoyam leaves, locally called *kontomire* in Ghana, are a popular vegetable for local sauces such as palaver sauce and egusi/agushi stew. It is also commonly consumed in Guinea and parts of Senegal, as a leaf sauce or as a vegetable side, and is referred to as "jaabere" in the local Pulaar dialect.

West Indies

Taro is called "dasheen", in contrast to the smaller variety of corms called "eddo," or tanya in the English speaking countries of the West Indies, and is cultivated and consumed as a staple crop in the region. There are differences amongst the roots mentioned above: taro or dasheen is mostly blue when cooked, tanya is white and very dry, and eddoes are small and very slimy.

In the Spanish speaking countries of the Spanish West Indies taro is called *ñame,* the Portuguese variant of which (*inhame*) is used in former Portuguese colonies where taro is still cultivated, including the Azores and Brazil. In Puerto Rico, it is sometimes called "malanga". In some countries, such as Trinidad and Tobago, St.Vincent, Dominica and the Grenadines, the leaves and stem of the dasheen, or taro, are most often cooked and pureed into a thick liquid called callaloo, which is served as a side dish similar to creamed spinach. Callaloo is sometimes prepared with crab legs, coconut milk, pumpkin and okra. It is usually served alongside rice or made into a soup along with various other roots.

Eddoe

Eddoe is a tropical vegetable, a variety of *Colocasia esculenta*, closely related to taro (dasheen), that is primarily used for its thickened stems (corms). It has smaller corms than taro, and in all but the best cultivars there is an acrid taste that requires careful cooking. The young leaves can also be cooked and eaten, but (unlike taro) they have a somewhat acrid taste.

Management

Eddoes appear to have been developed as a crop in China and Japan and introduced from there to the West Indies where they are sometimes called "Chinese eddoes". They grow best in rich loam soil with good drainage, but they can be grown in poorer soil, in drier climates, and in cooler temperatures than Taro.

Eddoes are also called malangas in Spanish-speaking areas, but that name is also used for other plants of the Araceae family, including tannia (*Xanthosoma* spp.). Eddoe is known as arvi or arbi in Urdu and Hindi and kochur mukhi in Bangla languages in South Asia.

In tropical Africa upland taro is often intercropped with other vegetables. In Ghana farmers normally intercrop taro with sugar cane and at times with maize before complete flooding of the field. In Hawaii taro is grown continuously for several years in flooded fields. After a few years the fields are dried and planted with tomato, cucumber or capsicum pepper before the field is returned again to flooded taro. In South-East Asia, upland taro is often grown on dykes of paddy fields.

Taro requires good soil fertility and adequate organic matter. Flooded taro requires greater quantities of fertilizer for maximum yields than upland taro. Potash is particularly important and the crop also has a relatively high calcium requirement. General recommendation for taro is 40–80 kg/ha N, 15–30 kg/ha P and 50–

100 kg/ha K. Some farmers in Ghana use organic manure (mainly from poultry) resulting in appreciable yields. Since the crop grows best in conditions where water is non-limiting, water management is a key factor. In situations with abundant water availability, closer spacing is advised; also when taro is intercropped it can be planted closer. Weeding is necessary until the canopy closes; in flooded cultivation growth of weeds is limited, and they can be easily controlled mechanically.

Propagation and Planting

Taro is propagated vegetatively. It is sometimes difficult to keep planting material in a healthy condition during the dry season or periods of drought. Essentially 4 types of planting material are used: side suckers growing from the main corm, small unmarketable cormels (60–150 g), corm pieces, and setts or 'huli', i.e. the apical 1–2 cm of the main corm with 15–20 cm of the leaf stalks attached. In Ghana planting is mainly by use of either young suckers or mature setts cut from harvested corms. Planting material must be taken from healthy plants. Cormels are planted at a depth of 5–7.5 cm. Planting on ridges facilitates harvesting. For upland cultivation in the Philippines, furrows 30 cm deep and 80 cm apart are prepared. In flooded culture fields are ploughed, puddled and carefully levelled. Planting is done after draining the field or into 2–5 cm of standing water. The planting distance is 50–80 cm in the row and 70–120 cm between rows. For breeding purposes taro can be propagated by seed.

Diseases and Pests

Taro blight (*Phytophthora colocasiae*) is a major wetland taro disease, causing purple to brown circular water-soaked lesions. It is the most devastating taro disease, particularly in the Pacific region, where it has caused considerable losses, e.g. in the Solomon Islands, which have resulted in the crop being partially replaced by sweet potato. This disease is partially controlled by use of copper- or phosphor-based fungicides, but spraying is tedious and costly. Increasing plant spacing or intercropping reduces losses. Resistance has been found in germplasm collections. Several species of *Pythium* cause taro soft rot, with wilting and chlorosis of leaves. Control is possible by use of healthy planting material, crop rotation and treating planting materials with fungicide. Sclerotium rot caused by *Sclerotium rolfsii* is characterized by stunting of the plant, rotting of corm and formation of many spherical sclerotia in the corm. Control is by soil drenching. In both flooded and upland taro, dark brown spots that appear in older leaves are caused by *Cladosporium colocasicola* and *Phyllosticta colocasiae*.

Dasheen mosaic virus (DsMV) and other viruses have been reported but are seldom serious. In the Pacific region the alomae virus disease causes serious damage. Symptoms start with a feathery mosaic on the leaves followed by crinkling and formation of outgrowths on the surface. Finally the entire plant becomes stunted and dies. Alomae disease is caused by the combined infestation by the taro large bacilliform virus (TLBV) and the taro small bacilliform virus (TSBV). Presence of TLBV only results in a milder form of the disease called 'bobone'. The viruses are transferred by a grasshopper and a mealy bug, respectively, but not by mechanical contact. In Hawaii, two diseases of unknown causative agents characterized by small, hardened portions in the lower third of the corm ('hard rot') and a soft rubbery corm which is low in starch content and exudes water when squeezed ('lohloli') are threatening. Attack by root-knot nematodes (*Meloidogyne* spp.) can result in considerable crop loss. Control is by treating planting material with water at 40°C for 50 minutes, by the use of disease-free material or by soil fumigation.

Insect pest on taro may cause serious damage. Damage by *Hercothrips indicus* (synonym: *Heliothrips indicus*) thrips is shown as a silvery discolouration of the leaves and can result in severe leaf shedding. Adults of taro beetles (*Papuana* spp., e.g. *Papuana huebneri* and *Papuana woodlarkiana*) tunnel in the corm up to the growing point. Young plants wilt and die but older plants usually recover. This pest is reported in the Pacific and South-East Asia, but not in Africa. It can be controlled by applying insecticide in the planting holes. Severe losses have been suffered by a number of countries growing the crop (Polynesia, Hawaii, the Caroline Islands and Samoa), as a result of leafhoppers. Biological control and insecticide dusting are effective control measures. Larvae of the sweet potato hawk moth (*Agrius convolvuli*) defoliating the plant reduce corm quality.

Ginger

Ginger or ginger root is the rhizome of the plant *Zingiber officinale*, consumed as a delicacy, medicine, or spice. It lends its name to its genus and family (Zingiberaceae). Other notable members of this plant family are turmeric, cardamom, and galangal. The distantly related dicots in the Asarum genus have the common name wild ginger because of their similar taste.

Figure: *Ginger*

Ginger cultivation began in South Asia and has since spread to East Africa and the Caribbean.

The English name ginger comes from French: *gingembre,* Old English: *gingifere,* Ultimately the origin is from the Dravidian word *inchi-ver*. The botanical term for root is *ver*, hence inchi root or inchi-ver.

Horticulture

Figure: *Ginger Plant with Flower - South India*

Ginger produces clusters of white and pink flower buds that bloom into yellow flowers. Because of its aesthetic appeal and the adaptation of the plant to warm climates, ginger is often used as landscaping around subtropical homes. It is a perennial reed-like plant with annual leafy stems, about a metre (3 to 4 feet) tall. Traditionally, the rhizome is gathered when the stalk withers; it is immediately scalded, or washed and scraped, to kill it and prevent sprouting. The fragrant perisperm of Zingiberaceae is used as sweetmeats by Bantu, also as a condiment and sialogogue.

Uses

Ginger produces a hot, fragrant kitchen spice. Young ginger rhizomes are juicy and fleshy with a very mild taste. They are often pickled in vinegar or sherry as a snack or just cooked as an ingredient in many dishes. They can also be steeped in boiling water to make ginger tea, to which honey is often added; sliced orange or lemon fruit may also be added. Ginger can also be made into candy, or ginger wine which has been made commercially since 1740. Mature ginger rhizomes are fibrous and nearly dry. The juice from old ginger roots is extremely potent and is often used as a spice in Indian recipes, and is a quintessential ingredient of Chinese, Korean, Japanese and many South Asian cuisines for flavouring dishes such as seafood or goat meat and vegetarian cuisine.

Ginger Acts as a Useful Food Preservative

Fresh ginger can be substituted for ground ginger at a ratio of 6 to 1, although the flavours of fresh and dried ginger are somewhat different. Powdered dry ginger root is typically used as a flavouring for recipes such as gingerbread, cookies, crackers and cakes, ginger ale, and ginger beer. Candied ginger, or crystallized ginger, is the root cooked in sugar until soft, and is a type of confectionery. Fresh ginger may be peeled before eating. For longer-term storage, the ginger can be placed in a plastic bag and refrigerated or frozen.

Regional Use

In Western cuisine, ginger is traditionally used mainly in sweet foods such as ginger ale, gingerbread, ginger snaps, parkin, ginger biscuits and speculaas. A ginger-flavoured liqueur called Canton is produced in Jarnac, France. Green ginger wine is a ginger-flavoured wine produced in the United Kingdom, traditionally sold in a green glass bottle. Ginger is also used as a spice added to hot coffee and tea.

In India and Pakistan, ginger is called *adrak* in sanskrit, Hindi, Punjabi and Urdu, *aad* in Maithili, *aadi* in Bhojpuri, *aada* in Assamese and Bengali, *Adu* in Gujarati, *hashi shunti* in *inji* in Tamil and Malayalam, *inguru* in Sinhalese, *alay* in Marathi, and *aduwa* in

Nepali. Fresh ginger is one of the main spices used for making pulse and lentil curries and other vegetable preparations. Fresh, as well as dried, ginger is used to spice tea and coffee, especially in winter. Ginger powder is also used in certain food preparations, particularly for pregnant or nursing women, the most popular one being *katlu* which is a mixture of gum resin, ghee, nuts, and sugar. Ginger is also consumed in candied and pickled form. In Bangladesh, ginger is finely chopped or ground into a paste to use as a base for chicken and meat dishes alongside onion and garlic.

In Burma, ginger is called *gyin*. It is widely used in cooking and as a main ingredient in traditional medicines. It is also consumed as a salad dish called *gyin-thot*, which consists of shredded ginger preserved in oil, and a variety of nuts and seeds. In Indonesia, a beverage called *wedang jahe* is made from ginger and palm sugar. Indonesians also use ground ginger root, called *jahe*, as a common ingredient in local recipes. In Malaysia, ginger is called *halia* and used in many kinds of dishes, especially a soup. In the Philippines it is brewed into a tea called *salabat*. In Vietnam, the fresh leaves, finely chopped, can also be added to shrimp-and-yam soup (*canh khoai mỡ*) as a top garnish and spice to add a much subtler flavour of ginger than the chopped root.

In China, sliced or whole ginger root is often paired with savory dishes such as fish, and chopped ginger root is commonly paired with meat, when it is cooked. However, candied ginger is sometimes a component of Chinese candy boxes, and a tisane can also be prepared from ginger.

In Japan, ginger is pickled to make beni shoga and gari or grated and used raw on tofu or noodles. It is also made into a candy called *shoga no sato zuke*. In the traditional Korean kimchi, ginger is either finely minced or just juiced in order to avoid the fibrous texture and added to the ingredients of the spicy paste just before the fermenting process.

In the Caribbean, ginger is a popular spice for cooking, and making drinks such as *sorrel*, a seasonal drink made during the Christmas season. Jamaicans make ginger beer both as a carbonated beverage and also fresh in their homes. Ginger tea is often made from fresh ginger, as well as the famous regional speciality Jamaican ginger cake.

On the island of Corfu, Greece, a traditional drink called (*tsitsibira*), a type of ginger beer, is made. The people of Corfu and the rest of the Ionian islands adopted the drink from the British, during the period of the United States of the Ionian Islands.

In Arabic, ginger is called *zanjabil,* and in some parts of the Middle East, *(uÜYon).* In Philippine cuisine it is called salabat and served in the relatively cold month of December. From its main ingredient ginger tea derives a flavour that is spicy and stimulating. Ginger, known as Adarak in Hindi, is used frequently in tea made in all parts of India as well.

Preliminary Research

Preliminary research indicates that nine compounds found in ginger may bind to human serotonin receptors which may influence gastrointestinal function. In a 2010 study, daily consumption of ginger was shown to help ease muscle pain associated with exercise by 25%. Ginger root supplement has been identified in one study to help reduce colon inflammation markers such as PGE2, thus indicating a measure that might affect colon cancer.

In limited studies, ginger was found to be more effective than placebo for treating nausea caused by seasickness, morning sickness and chemotherapy, although ginger was not found superior to placebo for pre-emptively treating post-operative nausea. Some studies advise against taking Ginger during pregnancy, this precautionary measure is based on studies suggesting that ginger is mutagenic, though some other studies have reported antimutagenic effects. Other preliminary studies showed that ginger may affect arthritis pain or have blood thinning and cholesterol lowering properties, but these effects remain unconfirmed.

Advanced glycation end-products are possibly associated in the development of diabetic cataract for which ginger was effective in preliminary studies, apparently by acting through antiglycating mechanisms. Zingerone may have activity against enterotoxigenic *Escherichia coli* in enterotoxin-induced diarrhea.

Chemistry

The characteristic odor and flavour of ginger is caused by a mixture of zingerone, shogaols and gingerols, volatile oils that compose one to three percent of the weight of fresh ginger. In laboratory animals, the gingerols increase the motility of the gastrointestinal tract and have analgesic, sedative, antipyretic and antibacterial properties. Ginger oil has been shown to prevent skin cancer in mice and a study at the University of Michigan demonstrated that gingerols can kill ovarian cancer cells. [6]-gingerol (1-[4'-hydroxy-3'-methoxyphenyl]-5-hydroxy-3-decanone) is the major pungent principle of ginger. The chemopreventive potentials of [6]-gingerol present a promising future alternative to expensive and toxic chemotherapeutic agents.

Ginger contains up to three percent of a fragrant essential oil whose main constituents are sesquiterpenoids, with (-)-zingiberene as the main component. Smaller amounts of other sesquiterpenoids (â-sesquiphellandrene, bisabolene and farnesene) and a small monoterpenoid fraction (â-phelladrene, cineol, and citral) have also been identified. The pungent taste of ginger is due to nonvolatile phenylpropanoid-derived compounds, particularly gingerols and shogaols, which form from gingerols when ginger is dried or cooked. Zingerone is also produced from gingerols during this process; this compound is less pungent and has a spicy-sweet aroma. Ginger is also a minor chemical irritant, and because of this was used as a horse suppository by pre-World War I mounted regiments for feaguing.

Ginger has a sialagogue action, stimulating the production of saliva, which makes swallowing easier.

Folk Medicine

The traditional medical form of ginger historically was called *Jamaica ginger*; it was classified as a stimulant and carminative and used frequently for dyspepsia, gastroparesis, slow motility symptoms, constipation, and colic. It was also frequently employed to disguise the taste of medicines.

Some studies indicate ginger may provide short-term relief of pregnancy-related nausea and vomiting. Studies are inconclusive about effects for other forms of nausea or in treating pain from rheumatoid arthritis, osteoarthritis, or joint and muscle injury. Side effects, mostly associated with powdered ginger, are gas, bloating, heartburn, and nausea. Tea brewed from ginger is a common folk remedy for colds. Ginger ale and ginger beer are also drunk as *stomach settlers* in countries where the beverages are made.

- In Burma, ginger and a local sweetener made from palm tree juice (*htan nyat*) are boiled together and taken to prevent the flu.
- In China, ginger is included in several traditional preparations. A drink made with sliced ginger cooked in water with brown sugar or a cola is used as a folk medicine for the common cold.

"Ginger eggs" (scrambled eggs with finely diced ginger root) is a common home remedy for coughing. The Chinese also make a kind of dried ginger candy that is fermented in plum juice and sugared, which is also commonly consumed to suppress coughing. Ginger has also been historically used to treat inflammation, which several scientific studies support, though one arthritis trial showed ginger to be no better than a placebo or ibuprofen for treatment of osteoarthritis.

- In Congo, ginger is crushed and mixed with mango tree sap to make tangawisi juice, which is considered a panacea.
- In India, ginger is applied as a paste to the temples to relieve headache, and consumed when suffering from the common cold. Ginger with lemon and black salt is also used for nausea.
- In Indonesia, ginger (*jahe* in Indonesian) is used as a herbal preparation to reduce fatigue, reducing "winds" in the blood, prevent and cure rheumatism and control poor dietary habits.
- In Nepal, ginger is called *aduwa,* and is widely grown and used throughout the country as a spice for vegetables, used medically to treat cold and also sometimes used to flavour tea.
- In the Philippines, ginger is known as *luya* and is used as a throat lozenge in traditional medicine to relieve sore throat. It is also brewed into a tea known as *salabat.*
- In the United States, ginger is used to prevent motion and morning sickness. It is recognised as safe by the Food and Drug Administration and is sold as an unregulated dietary supplement. Ginger water is also used to avoid heat cramps.
- In Peru, ginger is sliced in hot water as an infusion for stomach aches as *infusión de Kión.*
- In Japan it is purported to aid blood circulation. Scientific studies investigating these effects have been inconclusive.

Nutritional Information

Safety: If consumed in reasonable quantities, ginger has few negative side effects, and is on the FDA's "generally recognised as safe" list, though it does interact with some medications, including warfarin. Ginger is contraindicated in people suffering from gallstones, as it promotes the production of bile.

Allergic reactions to ginger generally result in a rash, and although generally recognised as safe, ginger can cause heartburn, bloating, gas, belching and nausea, particularly if taken in powdered form.

Unchewed fresh ginger may result in intestinal blockage, and individuals who have had ulcers, inflammatory bowel disease or blocked intestines may react badly to large quantities of fresh ginger. Ginger can also adversely affect individuals with gallstones. There are also suggestions that ginger may affect blood pressure, clotting, and heart rhythms.

Products in Taiwan made from Hebo Natural Products Limited of China contained ginger contaminated with DIBP, some 80,000 nutritional supplement capsules made with imported ginger powder were seized by the Public Health Department of Taiwan in June 2011.

Similar Ingredients

Myoga (*Zingiber mioga* Roscoe) appears in Japanese cuisine; the flower buds are the part eaten.

Another plant in the *Zingiberaceae* family, galangal, is used for similar purposes as ginger in Thai cuisine. Galangal is also called Thai ginger. Also referred to as galangal, fingerroot (*Boesenbergia rotunda*), or Chinese ginger or the Thai *krachai*, is used in cooking and medicine.

A dicotyledonous native species of eastern North America, *Asarum canadense*, is also known as "wild ginger", and its root has similar aromatic properties, but it is not related to true ginger. The plant also contains aristolochic acid, a carcinogenic compound.

Pests & Diseases – Beans

Diseases

Chocolate Spot (Botrytis Fabae, B. Cinerea)

- Symptoms appear as brown spots, which enlarge to give a more damaging aggressive phase in cool, wet or damp weather.
- Autumn sown beans are more likely to suffer yield losses, especially where the plant population is high and the crop becomes tall.

Leaf and Pod Spot (Ascochyta Fabae)

- This produces brown spots containing distinctive black fruiting bodies (pycnidia).
- Autumn sown beans are more prone to serious attacks especially in wet conditions.
- The disease is seed-borne, air-borne and splash dispersed.
- It is advised that farm-saved seed should be tested by PGRO.

- Infection can be transmitted from bean volunteers in neighbouring fields.

Downy Mildew (Peronospora Viciae)

- Mildew causes greyish-brown, felty growth on the under-surface of the leaves.
- It is prevalent on spring beans.
- Some varieties have resistance to the disease and ratings are given in the Recommended List of Varieties.

Rust (Uromyces Fabae)

- Rust is characterised by numerous small reddish-brown pustules on the leaves.
- It is more serious on spring beans and all varieties are susceptible.
- Most damage occurs if infection begins during flowering and pod set.

Sclerotinia (Sclerotinia Trifoliorum, S. Sclerotiorum)

- This disease occasionally infects winter beans in damp autumn weather.
- Plants develop a watery stem rot, which can spread between plants in dense stands.
- The related fungus, Sclerotinia sclerotiorum infects spring beans, peas, rape, linseed, lupins and a range of field vegetables.

Foot and Toot Rots (Fusarium Culmorum, F. Solani, Phoma Medicaginis var. Pinodella)

- The disease causes browning of the stem base and wilting of the leaves.
- It can occur on seedlings and on more mature plants.
- Foot and root rots in beans appear to be more sporadic than those which occur in peas, and the bean crop in general appears less sensitive to root rots than peas.

Pests

Pea and Bean Weevil (Sitona Lineatus)

- The adult is 4-5 mm long with a narrow body, short snout and angled antennae.
- The larva is white and legless with a brown head.

- Leaves of attacked plants show characteristic 'U' shaped notches around the edges.
- The main damage occurs as a result of the larvae feeding on the root nodules.
- The pest causes damage to spring sown beans if large numbers appear when plants are small.
- Autumn sown beans, although still prone to attack, are usually too advanced in growth for the weevil or the larvae to have any appreciable affect on yield.

Black Bean Aphid (Aphis Fabae) and Pea Aphid (Acyrthosiphon Pisum)

- Black bean aphid adults are 1-2 mm long, black with distinctive white or grey markings on the upper body.
- Pea aphid adults are 1-2 mm long, bright green, pear shaped with slightly dark legs and have distinctive red eyes.
- Black bean aphid can be very damaging to field and broad beans if colonies develop just prior to flowering.
- Spring-sown crops are usually more likely to suffer damaging attacks than autumn sown beans.
- As well as forming dense, smothering colonies on the upper part of the stem these and the less obvious pea aphid are able to transmit viruses which affect yield.

Stem Nematode (Ditylenchus Gigas, D. Dipsaci)

- Stem nematode is a slender, transparent, microscopic pest, invisible to the naked eye.
- The pest is seed-borne and can also infest soils, causing a problem for future bean crops.
- Plants appear distorted, stunted and twisted, and stems may show symptoms of reddish blistering.
- The nematode has become a major pest in field beans and can cause severe problems in wet seasons.
- Farm-saved seed from an infested stock that has been multiplied for several generations is a particular risk.
- Seed should be tested for nematode, and only clean stocks should be sown.

8

Classification and Identification of Diseases

Major Groups

Italian: Diverse in appearance, and biennial and annual in type, this group includes white, Romanesco, various green, purple, brown and yellow cultivars. This type is the ancestral form from which the others were derived.

Northwest European Biennial: Used in Europe for winter and early spring harvest, this was developed in France in the 19th century, and includes the old cultivars Roscoff and Angers.

Northern European Annuals: Used in Europe and North America for summer and fall harvest, it was developed in Germany in the 18th century, and includes the old cultivars Erfurt and Snowball.

Asian: A tropical cauliflower used in China and India, it was developed in India during the 19th century from the now-abandoned Cornish type, and includes old varieties. Early Patna and Early Benaras.

Varieties

There are hundreds of historic and current commercial varieties used around the world. A comprehensive list of about 80 North American varieties is maintained at North Carolina State University.

Colours

White: White cauliflower is the most common colour of cauliflower.

Orange: Orange cauliflower (*B. oleracea* L. var. *botrytis*) contains 25% more vitamin A than white varieties. This trait came from a natural mutant found in a cauliflower field in Canada. Cultivars include 'Cheddar' and 'Orange Bouquet'.

Green: Green cauliflower, of the *B. oleracea* botrytis group, is sometimes called broccoflower. It is available both with the normal curd shape and a variant spiky curd called Romanesco broccoli. Both types have been commercially available in the U.S. and Europe since the early 1990s. Green-curded varieties include 'Alverda', 'Green Goddess' and 'Vorda'. Romanesco varieties include 'Minaret' and 'Veronica'.

Purple: The purple colour in this cauliflower is caused by the presence of the antioxidant group anthocyanins, which can also be found in red cabbage and red wine. Varieties include 'Graffiti' and 'Purple Cape'. In Great Britain and southern Italy, a broccoli with tiny flower buds is sold as a vegetable under the name "purple cauliflower". It is not the same as standard cauliflower with a purple curd.

Figure: *White cauliflower on display in a supermarket in Melbourne, Australia*

Nutrition

Cauliflower is low in fat, low in carbohydrates but high in dietary fibre, folate, water, and vitamin C, possessing a high nutritional density.

Cauliflower contains several phytochemicals, common in the cabbage family, that may be beneficial to human health.

- Sulforaphane, a compound released when cauliflower is chopped or chewed, may protect against cancer.
- Other glucosinolates
- Carotenoids
- Indole-3-carbinol, a chemical that enhances DNA repair, and acts as an estrogen antagonist, slowing the growth of cancer cells.

Boiling reduces the levels of these compounds, with losses of 20–30% after five minutes, 40–50% after ten minutes, and 75% after thirty minutes. However, other preparation methods, such as steaming, microwaving, and stir frying, have no significant effect on the compounds.

A high intake of cauliflower has been associated with reduced risk of aggressive prostate cancer.

Cooking

Cauliflower can be roasted, boiled, fried, steamed, or eaten raw. Steaming or microwaving better preserves anticancer compounds than boiling. When cooking, the outer leaves and thick stalks are removed, leaving only the florets. The leaves are also edible, but are most often discarded. The florets should be broken into similar-sized pieces so they are cooked evenly. After eight minutes of steaming, or five minutes of boiling, the florets should be soft, but not mushy (depending on size). Stirring while cooking can break the florets into smaller, uneven pieces.

Low carbohydrate dieters can use cauliflower as a reasonable substitute for potatoes or rice; while they can produce a similar texture, or mouth feel, they lack the starch of the originals.

Fractal Dimension

Figure: *Fractal pattern of Romanesco broccoli, a variant form of cauliflower*

Cauliflower has been noticed by mathematicians for its distinct fractal dimension, predicted to be about 2.8.

History

The first reliable reference to cauliflower is found in the writings of the Arab Muslim scientists Ibn al-'Awwam and Ibn al-Baitar, in the 12th and 13th centuries.

Propagation

Cauliflower is a cool season crop and grows best in well draining, organic soil at a pH of 6.5. It should be planted after the last frost as sub-freezing temperatures are likely to damage the plant. It is most successfully grown from transplants started in the glasshouse. Transplants should be approximately 6 weeks old before setting in the field. Plant spacing will greatly affect the final size of the cauliflower head and the plants should be spaced at least 46 cm (18 in) apart leaving approximately 75 cm (30 in) between rows.

Cauliflower heads can be discoloured by sunlight and when the cauliflower head has reached 2.5–5.0 cm (1–2 in) the plant requires 'blanching', a process that ensures that the head remains white. This is achieved by gathering the longest outer leaves together at the head and tying them with twine.

Cabbage

Cabbage (*Brassica oleracea* or variants) is a leafy green biennial, grown as an annual vegetable for its dense-leaved heads. Closely related to other cole crops, such as broccoli, cauliflower, and brussels sprouts, it descends from *B. oleracea* var. *oleracea*, a wild field cabbage. Cabbage heads generally range from 1 to 8 pounds (0.5 to 4 kg), and can be green, purple and white. Smooth-leafed firm-headed green cabbages are the most common, with smooth-leafed red and crinkle-leafed savoy cabbages of both colours seen more rarely.

It is difficult to trace the exact history of cabbage, but it was most likely domesticated somewhere in Europe before 1000 BC. By the Middle Ages, it was a prominent part of European cuisine, although savoys were not developed until the 16th century. Cabbage heads are generally picked during the first year of the plants' life cycles, but those intended for seed are allowed to grow a second year, and must be kept separated from other cole crops to prevent cross pollination. Cabbage is prone to several nutrient deficiencies, as well as multiple pests, bacteria and fungal diseases.

The Food and Agriculture Organisation of the United Nations (FAO) reports that world production of cabbage and other brassicas (these plants are combined by the FAO for reporting purposes) for calendar year 2010 was almost 58,000,000 metric tons (57,000,000 long tons; 64,000,000 short tons). Almost half were grown in China. Cabbages are prepared in many different ways for eating, although pickling, in dishes such as sauerkraut, is the most popular. Cabbage is a good source of beta-carotene, vitamin C, and fibre. Cabbage when contaminated is sometimes a source of food-borne illness in humans.

Description

Cabbage seedlings have a thin taproot and cordate (heart-shaped) cotyledons. The first leaves produced are ovate (egg-shaped) with a lobed petiole. Plants are 40–60 cm (16–24 in) tall in their first year at the mature vegetative stage, and 1.5–2.0 m (4.9–6.6 ft) tall when flowering in the second year. Heads average between 1 and 8 pounds (0.5 and 4 kg), with earlier varieties producing smaller heads. Most cabbages have thick, alternating leaves, with margins that range from wavy or lobed to highly dissected; some varieties have a waxy bloom on the leaves. Plants have root systems that are fibrous and shallow. About 90 percent of the root mass is in the upper 20–30 cm (7.9–12 in) of soil, although some lateral roots can penetrate up to 2 m (6.6 ft) deep.

The inflorescence is an unbranched and indeterminate terminal raceme measuring 50–100 cm (20–39 in), with flowers that are yellow or white. Each flower has four petals set in a perpendicular pattern, as well as four sepals, six stamens, and a superior ovary that is two-celled and contains a single stigma and style. Two of the six stamens have shorter filaments. The fruit is a silique that opens at maturity through dehiscence to reveal brown or black seeds that are small and round in shape. Self-pollination is impossible, and plants are cross-pollinated by insects. The initial leaves form a rosette shape comprising 7 to 15 leaves, each measuring 25–35 cm (9.8–14 in) by 20–30 cm (7.9–12 in); after this, leaves with shorter petioles develop and heads form through the leaves cupping inward.

Many shapes, colours and leaf textures are found in various cabbage varieties. Leaf types are generally divided between crinkled-leaf, loose-head savoys and smooth-leaf firm-head cabbages, while the colour spectrum includes white and a range of greens and purples. Oblate, round and pointed shapes are found.

Cabbage has been selectively bred for head weight and morphological characteristics, frost hardiness, fast growth and storage ability. The appearance of the cabbage head has been given importance in selective breeding, with varieties being chosen for shape, colour, firmness and other physical characteristics. Breeding objectives are now focused on increasing resistance to various insects and diseases and improving the nutritional content of cabbage. Scientific research into the genetic modification of *B. oleracea* crops, including cabbage, has included European Union and United States explorations of greater insect and herbicide resistance. However, genetically modified *B. oleracea* crops are not currently used in commercial agriculture.

History

Although cabbage has an extensive history, it is difficult to trace its exact origins owing to the many varieties of leafy greens classified as "brassicas". The wild ancestor of cabbage was originally found in Britain and continental Europe. It was probably domesticated later in history than Near East crops such as peas and summer wheat. Because of the wide range of crops developed from the wild *B. oleracea*, multiple domestications of cabbage may have occurred in the same time period throughout Europe.

Nonheading cabbages and kale were probably the first to be domesticated, sometime before 1000 BC, and the Greeks and Romans had some variety of cabbage, although whether it was more closely related to today's cabbage or to one of the other *Brassica* crops is unknown. The Greeks and Romans claimed medicinal usages for their cabbage variety that included relief from gout, headaches and the symptoms of poisonous mushroom ingestion. Rounded cabbages made a definite appearance during the Dark Ages, and illustrations show their prominence in the cuisine of the Middle Ages. French naturalist Jean Ruel made what is considered the first mention of head cabbage in his 1536 botanical treatise *De Natura Stirpium*, referring to it as *capucos coles* ("head-stem"). Cabbages spread from Europe into Mesopotamia and Egypt, and later followed trade routes throughout Asia and the Americas.

During the 16th century, German gardeners developed the savoy cabbage. During the 17th and 18th centuries, cabbage was a food staple in such countries as Germany, England, Ireland and Russia, and pickled cabbage was frequently seen. Sauerkraut was used by Dutch sailors to prevent scurvy during long ship voyages. Jacques Cartier first brought cabbage to the Americas in 1541–42, and it was probably planted in what is now the United States by the early colonists, despite the lack of written evidence of its existence there until the mid-17th century. By the 18th century, it was commonly planted by both colonists and native American Indians. Cabbage seeds travelled to Australia in 1788 with the First Fleet, and were planted the same year on Norfolk Island. It became a favourite vegetable of Australians by the 1830s and was frequently seen at the Sydney Markets.

There are several Guinness Book of World Records related to cabbage. These include the heaviest cabbage, at 57.61 kilograms (127.0 lb), heaviest red cabbage, at 19.05 kilograms (42.0 lb), longest cabbage roll, at 15.37 metres (50.4 ft), and the largest cabbage dish, at 925.4 kilograms (2,040 lb). In 2012, Scott Robb of Palmer, Alaska, broke the world record for heaviest cabbage at 138.25 pounds (62.71 kg).

Cultivation

Figure: *A cabbage field in Dojran Municipality, Republic of Macedonia*

Cabbage is generally grown for its densely leaved heads, produced during the first year of its biennial cycle. Plants perform best when grown in well-drained soil in a location that receives full sun. Different varieties prefer different soil types, ranging from lighter sand to heavier clay, but all prefer fertile ground with a pH of between 6.0 and 6.8.

Temperatures of between 40 and 75 °F (4 and 24 °C) prompt the best growth, and extended periods of higher or lower temperatures may result in premature bolting (flowering). Plants are generally started in protected locations early in the growing season before being transplanted outside, although some are seeded directly into the ground from which they will be harvested.

Seedlings typically emerge in about 4–6 days from seeds planted 0.5 inches deep at a soil temperature between 68 and 86 °F (20 and 30 °C). Growers normally place plants 12 to 24 inches (30 to 61 cm) apart. Some varieties of cabbage have been developed for ornamental use, these are generally called “flowering cabbage”. They do not produce heads and feature purple or green outer leaves surrounding an inner grouping of smaller leaves in white, red, or pink. Early varieties of cabbage take about 70 days from planting to reach maturity, while late varieties take about 120 days. Cabbages are mature when they are firm and solid to the touch. They are harvested by cutting the stalk just below the bottom leaves with a knife. The outer leaves are trimmed, and any diseased, damaged, or necrotic leaves are removed. When being grown for seed, cabbages must be isolated from other *B. oleracea* subspecies, including the wild varieties, by .5 to 1 mile (0.80 to 1.6 km) to prevent cross pollination. Other *Brassica* species, such as *B. rapa*, *B. juncea*, *B. nigra*, *B. napus* and *Raphanus sativus*, do not easily cross-pollinate.

Cultivars

Figure: *White cabbage*

There are several cultivars of cabbage, each including many varieties:

- Savoy – Characterized by crimped or curly leaves, mild flavour and tender texture
- Spring Greens – Loose-headed, commonly sliced and steamed
- Green – Light to dark green, slightly pointed heads. This is the most commonly grown cultivar.
- Red – Smooth red leaves, often used for pickling or stewing
- White (also called Dutch) – Smooth, pale green leaves

Some sources only delineate three cultivars: savoy, red and white, with spring greens and green cabbage being subsumed into the latter.

Cultivation Problems

Due to its high level of nutrient requirements, cabbage is prone to nutrient deficiencies, including boron, calcium, phosphorus and potassium. There are several physiological disorders that can affect the postharvest appearance of cabbage. Internal tip burn occurs when the margins of inside leaves turn brown, but the outer leaves look normal. Necrotic spot is where there are oval sunken spots a few millimetres across that are often grouped around the midrib. In pepper spot, tiny black spots occur on the areas between the veins, which can increase during storage.

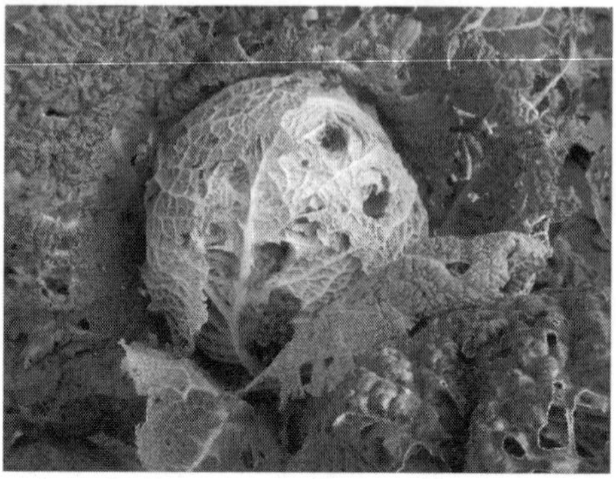

Figure: *Cabbage moth damage to a savoy*

Fungal diseases include wirestem, which causes weak or dying transplants; *Fusarium* yellows, which cause stunted and twisted plants with yellow leaves; and blackleg, which causes sunken areas on stems and gray-brown spotted leaves. The fungi *Alternaria brassicae* and *A. brassicicola* cause dark leaf spots in affected plants. They are both seedborne and airborne, and typically propagate from spores in infected plant debris left on the soil surface for up to twelve weeks after harvest. *Rhizoctonia solani* causes the post-emergence disease wirestem, resulting in killed seedlings ("damping-off"), root rot, or stunted growth and smaller heads.

Bacterial diseases include black rot, caused by *Xanthomonas campestris*, which causes cholorotic and necrotic lesions that start at the leaf margins, and wilting of plants. Clubroot, caused by the soilborne slime mold-like organism *Plasmodiophora brassicae*, results in swollen, club-like roots.

A parasitic disease downy mildew, caused by the oomycete *Peronospora parasitica*, produces pale leaves with white, brownish or olive mildew on the lower leaf surfaces; this is often confused with the fungal powdery mildew.

Pests include root-knot nematodes and cabbage maggots, which cause stunted and wilted plants with yellow leaves; aphids, which cause stunted plants with curled and yellow leaves; harlequin bugs, which cause white and yellow leaves; thrips, which cause leaves with white-bronze spots; striped flea beetles, which riddle leaves with small holes; and caterpillars, which leave behind large, ragged holes in leaves.

The caterpillar stage of the "small cabbage white butterfly" (*Pieris rapae*), commonly known as the "imported cabbage worm", is a major cabbage pest in most countries. The large white butterfly (*Pieris brassicae*) is prevalent in eastern European countries.

The diamondback moth (*Plutella xylostella*) and the cabbage moth (*Mamestra brassicae*) thrive in the higher summer temperatures of continental Europe, where they cause considerable damage to cabbage crops.

Destructive soil insects include the cabbage root fly (*Delia radicum*) and the cabbage maggot (*Hylemya brassicae*), whose larvae can burrow into the part of plant consumed by humans.

Planting near other members of the cabbage family, or where these plants have been placed in previous years, can prompt the spread of pests and disease. Excessive water and excessive heat can also cause cultivation problems.

Production

The Food and Agriculture Organisation of the United Nations (FAO) reports that world production of cabbage and other brassicas (these plants being combined by the FAO for reporting purposes) for calendar year 2010 was 57,966,986 metric tons (57,051,486 long tons; 63,897,664 short tons). This was primarily grown in China (43 percent) and India (11 percent).

Table: *Top Ten Cabbage and Other Brassica Producers — 2010*

Country	***Production (tonnes)***	***Source***
People's Republic of China	25,156,578	FAO estimate
India	6,356,800	official figure
Russia	2,732,510	official figure
Japan	2,247,700	official figure
South Korea	2,035,700	official figure
Ukraine	1,497,400	official figure
Indonesia	1,384,660	official figure
Poland	1,141,200	official figure
United States	1,034,050	official figure
Romania	983,648	official figure
World	57,966,986	aggregate

Cabbages sold for fresh market are generally smaller, and different varieties are used for those sold immediately upon harvest and those stored before sale. Those used for processing, especially sauerkraut, are larger and have a lower percentage of water. Both hand and mechanical harvesting are used, with hand-harvesting generally used for cabbages destined for fresh market sales. Field packing of hand-harvested cabbages is practiced in commercial-scale operations. In large operations, after harvest, heads are trimmed, sorted and packed, then vacuum cooled. Cabbage can be stored the longest at between 30 to 36 °F (-1 to 2 °C) with a humidity of 90–100 percent; these conditions will result in up to six months of longevity. When stored under less ideal conditions, cabbage can still last up to four months.

Culinary Use

Cabbage is used in many ways, ranging from eating raw and simple steaming to pickling, stewing, sauteing or braising. Pickling is one of the most popular ways of preserving cabbage, creating dishes such as sauerkraut and kimchee, although kimchee is more often made from Chinese cabbage (*B. rapa*). Savoy cabbages are usually used in salads, while smooth-leaf types are utilised for both fresh market sales and processing. Bean curd and cabbage is a staple of Chinese cooking, while the British dish bubble and squeak is made primarily with salt beef and boiled cabbage. Cabbage is used extensively in Polish cuisine. It is one of the main food crops, and sauerkraut is a frequent dish, as well as being used to stuff other dishes such as golabki (stuffed cabbage) and pierogi (filled pasta). Other eastern European countries, such as Hungary and Romania, also have traditional dishes that feature cabbage as a main ingredient. In the United States, cabbage is used primarily for the production of coleslaw, followed by fresh market use and sauerkraut production. Cabbage consumption varies widely around the world, with the Russians eating the largest amount in Europe, at 20 kilograms (44 lb) per capita, while Belgians consume 4.7 kilograms (10 lb), the Dutch 4.0 kilograms (8.8 lb), Americans 3.9 kilograms (8.6 lb) and the Spaniards 1.9 kilograms (4.2 lb).

The characteristic flavour of cabbage is caused by glucosinolates, a class of sulfur-containing glucosides. Although found throughout the plant, these compounds are concentrated in the highest quantities in the seeds; lesser quantities are found in young vegetative tissue, and they decrease as the tissue ages. Cooked cabbage is often criticized for its pungent, unpleasant odor and taste. These develop when cabbage is overcooked and a hydrogen sulfide gas is produced.

Nutrition and Health

Cabbage is a good source of beta-carotene, vitamin C and fibre. It is a cruciferous vegetable, and has been shown to reduce the risk of some cancers, especially those in the colourectal group. This is possibly due to the glucosinolates found in cole crops, which serve as metabolic detoxicants, or due to the sulphoraphane content, also responsible for metabolic anti-carcinogenic activities. Purple cabbage also contains anthocyanins, which in other vegetables have been proven to have anti-carcinogenic properties. Along with other cole crops, cabbage is a source of indole-3-carbinol, a chemical that boosts DNA repair in cells and appears to block the growth of cancer cells. Research suggests that boiling these vegetables reduces their anti-carcinogenic properties.

Food-Borne Illness

Cabbage has been linked to outbreaks of some food-borne illnesses, including *Listeria monocytogenes* and *Clostridium botulinum*. The latter toxin has been traced to pre-made, packaged coleslaw mixes, while the spores were found on whole cabbages that were otherwise acceptable in appearance. *Shigella* species are able to survive in shredded cabbage. Two outbreaks of *E. coli* in the United States have been linked to cabbage consumption. Biological risk assessments have concluded that there is the potential for further outbreaks linked to cabbage, due to contamination at many stages of the growing, harvesting and packaging processes. Contaminants from water, humans, animals and soil have the potential to be transferred to cabbage, and from there to the end consumer.

Medicinal Lore

In addition to its usual purpose as an edible vegetable, cabbage has been used historically as a medicinal herb. Ancient Greeks recommended using the juice of cabbage as an antidote for mushroom poisoning. The ancient Roman nobleman Pliny the Elder described both culinary and medicinal properties of the vegetable, recommending it for drunkenness–both preventatively to prevent the effects of alcohol, and to cure hangovers. This traditional usage persisted in European literature until the mid-20th century. In Cato the Elder's work *De Agri Cultura* ("On Agriculture"), he suggested that women could prevent diseases in their private parts by bathing in urine obtained from those who had frequently eaten cabbage. The cooling properties of the leaves were used in Britain as compresses for ulcers and breast abscesses, and as a treatment for trench foot in World War I. Other medicinal uses recorded in Europe folk medicine include treatments for rheumatism, sore throat, hoarseness, colic, and melancholy.

In the United States, cabbage has been used as a hangover cure, to treat abscesses, to prevent sunstroke, or to cool body parts affected by fevers. The leaves have also been used to sooth sore feet, and, tied around the neck of children, to relieve croup. Mashed cabbage and cabbage juice have been used in poultices to remove boils and treat warts, pneumonia, appendicitis, and ulcers.

Mint

Mentha (also known as Mint, from Greek *míntha*, Linear B *mi-ta*) is a genus of plants in the family Lamiaceae (mint family). The species are not clearly distinct and estimates of the number of species varies from 13 to 18. Hybridization between some of the species occurs naturally. Many other hybrids as well as numerous cultivars are known in cultivation. The genus has a subcosmopolitan distribution across Europe, Africa, Asia, Australia and North America.

Figure: Mentha longifolia

Mints are aromatic, almost exclusively perennial, rarely annual, herbs. They have wide-spreading underground and overground stolons and erect, square, branched stems. The leaves are arranged in opposite pairs, from oblong to lanceolate, often downy, and with a serrate margin. Leaf colours range from dark green and gray-green to purple, blue, and sometimes pale yellow. The flowers are white to purple and produced in false whorls called verticillasters. The corolla is two-lipped with four subequal lobes, the upper lobe usually the largest. The fruit is a small, dry capsule containing one to four seeds.

While the species that make up the Mentha genus are widely distributed and can be found in many environments, most Mentha grow best in wet environments and moist soils. Mints will grow 10–120 cm tall and can spread over an indeterminate area. Due to their tendency to spread unchecked, mints are considered invasive.

Species

The list below includes all of the taxa that have been recognised as species in recent works on *Mentha*. No author has recognised all of them. As with all biological classifications of plants, this list can go out of date at a moment's notice. Common names are also given for species that have them. Synonyms, along with cultivars and varieties are given in articles on the species.

Cultivation

Figure: Mentha x gracilis *and* M. rotundifolia. *The steel ring is to control the spread of the plant.*

All mints prefer, and thrive near pools of water, lakes, rivers, cool, moist spots in partial shade. In general, mints tolerate a wide range of conditions, and can also be grown in full sun.

They are fast growing, extending their reach along surfaces through a network of runners. Due to their speedy growth, one plant of each desired mint, along with a little care, will provide more than enough mint for home use. Some mint species are more invasive than others. Even with the less invasive mints, care should be taken when mixing any mint with any other plants, lest the mint take over. To control mints in an open environment, mints should be planted in deep, bottomless containers sunk in the ground, or planted above ground in tubs and barrels.

Some mints can be propagated by seed. Growth from seed can be an unreliable method for raising mint for two reasons: mint seeds are highly variable - one might not end up with what one presupposed was planted; and some mint varieties are sterile. It is more effective to take and plant cuttings from the runners of healthy mints.

The most common and popular mints for cultivation are peppermint (*Mentha* × *piperita*), spearmint (*Mentha spicata*), and (more recently) apple mint (*Mentha suaveolens*).

Mints are supposed to make good companion plants, repelling pest insects and attracting beneficial ones. Mints are susceptible to whitefly and aphids.

Harvesting of mint leaves can be done at any time. Fresh mint leaves should be used immediately or stored up to a couple of days in plastic bags within a refrigerator. Optionally, mint can be frozen in ice cube trays. Dried mint leaves should be stored in an airtight container placed in a cool, dark, dry area.

Uses Culinary

The leaf, fresh or dried, is the culinary source of mint. Fresh mint is usually preferred over dried mint when storage of the mint is not a problem. The leaves have a warm, fresh, aromatic, sweet flavour with a cool aftertaste. Mint leaves are used in teas, beverages, jellies, syrups, candies, and ice creams. In Middle Eastern cuisine, mint is used on lamb dishes, while in British cuisine and American cuisine, mint sauce and mint jelly are used, respectively.

Mint is a necessary ingredient in Touareg tea, a popular tea in northern African and Arab countries.

Figure: *Limonana (mint lemonade) served in Damascus, Syria.*

Alcoholic drinks sometimes feature mint for flavour or garnish, such as the mint julep and the mojito. Crème de menthe is a mint-flavoured liqueur used in drinks such as the grasshopper.

Mint essential oil and menthol are extensively used as flavourings in breath fresheners, drinks, antiseptic mouth rinses, toothpaste, chewing gum, desserts, and candies. The substances that give the mints their characteristic aromas and flavours are menthol (the main aroma of Peppermint and Japanese Peppermint) and pulegone (in Pennyroyal and Corsican Mint). The compound primarily responsible for the aroma and flavour of spearmint is *R*-carvone.

Mints are used as food plants by the larvae of some Lepidoptera species including Buff Ermine.

Medicinal and Cosmetic

Mint was originally used as a medicinal herb to treat stomach ache and chest pains, and it is commonly used in the form of tea as a home remedy to help alleviate stomach pain. In Rome, Pliny recommended that a wreath of mint was a good thing for students to wear since it was thought to "exhilarate their minds". During the Middle Ages, powdered mint leaves were used to whiten teeth. Mint tea is a diuretic. A common use is as an antipruritic, especially in insect bite treatments (often along with camphor). The strong, sharp flavour and scent of mint is sometimes used as a mild decongestant for illnesses such as the common cold. Mint is also used in some shampoo products.

Menthol from mint essential oil (40–90%) is an ingredient of many cosmetics and some perfumes. Menthol and mint essential oil are also much used in medicine as a component of many drugs, and are very popular in aromatherapy. Menthol is also used in cigarettes as an additive, because it blocks out the bitter taste of tobacco and soothes the throat.

Although it is used to treat many symptoms, mint can also cause allergic reactions to some people. Although rare, these can induce painful symptoms including abdominal cramps and diarrhoea, headaches, tingling or numbing around the mouth, nasal congestion, clogging of the sinuses, nausea, etc. It can be a reaction to Salycilates or Linalol contained in the mint, or to some of the proteins inside the plant. It is possible to have high intolerance to one type of mint such as spearmint yet have no reaction to other types such as peppermint and menthol, and the symptoms may get worse over time. Because it is uncommon, the people who suffer from mint allergies can find it hard to deal with it on a daily basis. Most of the problems arise from the need to find special toothpaste or dental products, but can also be reactions to the smell of somebody consuming mint candy or gum in a public or work place. In some cases it can be a minor disturbance but sometimes a severe reaction: because it is potent, inhaling the mint can trigger by itself breathing problems, nausea and dizziness separately or all together. Depending on whether the allergy is to a chemical or a protein inside the plant, it may or may not extend to other Lamiaceae.

iFood.tv claims that the American Allergy and asthma Foundation has tracked an increasing amount of mint allergies as with allergies in general.

Insecticides

Mint oil is also used as an environmentally friendly insecticide for its ability to kill some common pests like wasps, hornets, ants and cockroaches.

Room Scent and Aromatherapy

Known in Greek mythology as the herb of hospitality, one of mint's first known uses in Europe was as a room deodorizer. The herb was strewn across floors to cover the smell of the hard-packed soil. Stepping on the mint helped to spread its scent through the room. Today, it is more commonly used for aromatherapy through the use of essential oils.

Diseases

Etymology of "Mint"

Figure: *An example of mint leaves*

Mint descends from the Latin word *mentha,* which is rooted in the Greek word *minthe,* personified in Greek mythology as Minthe, a nymph who was transformed into a mint plant. The word itself probably derives from a now extinct pre-Greek language.

Mint leaves, without a qualifier like *peppermint* or *apple mint,* generally refers to spearmint leaves.

In Spain, Central and South America, mint is known as *hierba buena* (literally, "good herb"). In Lusophone countries, especially in Portugal, mint species are popularly known as *Hortelã.* In many Indo-Aryan languages, it is called *Pudîna.*

The taxonomic family Lamiaceae is known as the *mint family.* It includes many other aromatic herbs, including most of the more common cooking herbs, including basil, rosemary, sage, oregano, and catnip.

As an English colloquial term, any small mint-flavoured confectionery item can be called a mint.

In common usage, other plants with fragrant leaves may be called "mint" although they are not in the Mint Family.

- Vietnamese Mint, commonly used in Southeast Asian cuisine is *Persicaria odorata* in the family Polygonaceae, collectively known as smartweeds or pinkweeds.
- "Mexican mint marigold" is *Tagetes lucida* in the sunflower family (Asteraceae).

9

Plant Growth Regulators

Plant hormones (also known as phytohormones) are chemicals that regulate plant growth, which, in the UK, are termed 'plant growth substances'.

Plant hormones are signal molecules produced within the plant, and occur in extremely low concentrations. Hormones regulate cellular processes in targeted cells locally and, when moved to other locations, in other locations of the plant. Hormones also determine the formation of flowers, stems, leaves, the shedding of leaves, and the development and ripening of fruit. Plants, unlike animals, lack glands that produce and secrete hormones. Instead, each cell is capable of producing hormones. Plant hormones shape the plant, affecting seed growth, time of flowering, the sex of flowers, senescence of leaves, and fruits. They affect which tissues grow upward and which grow downward, leaf formation and stem growth, fruit development and ripening, plant longevity, and even plant death. Hormones are vital to plant growth, and, lacking them, plants would be mostly a mass of undifferentiated cells. So they are also known as growth factors or growth hormones. The term 'Phytohormone' was coined by Thimann in 1948.

Characteristics

The word hormone is derived from Greek, meaning *set in motion*. Plant hormones affect gene expression and transcription levels, cellular division, and growth. They are naturally produced within plants, though very similar chemicals are produced by fungi and bacteria that can also affect plant growth. A large number of related chemical compounds are synthesized by humans. They are used to regulate the growth of cultivated plants, weeds, and in vitro-grown plants and plant cells; these manmade compounds are called Plant Growth Regulators or PGRs for short. Early in the study of plant hormones, "phytohormone" was the commonly used term, but its use is less widely applied now.

Plant hormones are not nutrients, but chemicals that in small amounts promote and influence the growth, development, and differentiation of cells and tissues. The biosynthesis of plant hormones within plant tissues is often diffuse and not always localized. Plants lack glands to produce and store hormones, because, unlike animals — which have two circulatory systems (lymphatic and cardiovascular) powered by a heart that moves fluids around the body — plants use more passive means to move chemicals around the plant. Plants utilise simple chemicals as hormones, which move more easily through the plant's tissues. They are often produced and used on a local basis within the plant body. Plant cells produce hormones that affect even different regions of the cell producing the hormone.

Hormones are transported within the plant by utilising four types of movements. For localized movement, cytoplasmic streaming within cells and slow diffusion of ions and molecules between cells are utilised. Vascular tissues are used to move hormones from one part of the plant to another; these include sieve tubes or phloem that move sugars from the leaves to the roots and flowers, and xylem that moves water and mineral solutes from the roots to the foliage.

Not all plant cells respond to hormones, but those cells that do are programmed to respond at specific points in their growth cycle. The greatest effects occur at specific stages during the cell's life, with diminished effects occurring before or after this period. Plants need hormones at very specific times during plant growth and at specific locations. They also need to disengage the effects that hormones have when they are no longer needed. The production of hormones occurs very often at sites of active growth within the meristems, before cells have fully differentiated. After production, they are sometimes moved to other parts of the plant, where they cause an immediate effect; or they can be stored in cells to be released later. Plants use different pathways to regulate internal hormone quantities and moderate their effects; they can regulate the amount of chemicals used to biosynthesize hormones. They can store them in cells, inactivate them, or cannibalise already-formed hormones by conjugating them with carbohydrates, amino acids, or peptides. Plants can also break down hormones chemically, effectively destroying them. Plant hormones frequently regulate the concentrations of other plant hormones. Plants also move hormones around the plant diluting their concentrations.

The concentration of hormones required for plant responses are very low (10^{-6} to 10^{-5} mol/L). Because of these low concentrations, it has been very difficult to study plant hormones, and only since the late 1970s have scientists been able to start piecing together their effects and relationships to plant physiology. Much of the early work on plant hormones involved studying plants that were genetically deficient in one or involved the use of tissue-cultured plants grown *in vitro* that were subjected to differing ratios of hormones, and the resultant growth compared. The earliest scientific observation and study dates to the 1880s; the determination and observation of plant hormones and their identification was spread-out over the next 70 years.

Classes of Plant Hormones

In general, it is accepted that there are five major classes of plant hormones, some of which are made up of many different chemicals that can vary in structure from one plant to the next. The chemicals are each grouped together into one of these classes based on their structural similarities and on their effects on plant physiology. Other plant hormones and growth regulators are not easily grouped into these classes; they exist naturally or are synthesized by humans or other organisms, including chemicals that inhibit plant growth or interrupt the physiological processes within plants. Each class has positive as well as inhibitory functions, and most often work in tandem with each other, with varying ratios of one or more interplaying to affect growth regulation.

The five major classes are:

Abscisic Acid

Abscisic acid (also called ABA) is one of the most important plant growth regulators. It was discovered and researched under two different names before its chemical properties were fully known, it was called *dormin* and *abscicin II*. Once it was determined that the two compounds are the same, it was named abscisic acid. The name "abscisic acid" was given because it was found in high concentrations in newly abscissed or freshly fallen leaves.

This class of PGR is composed of one chemical compound normally produced in the leaves of plants, originating from chloroplasts, especially when plants are under stress. In general, it acts as an inhibitory chemical compound that affects bud growth, and seed and bud dormancy. It mediates changes within the apical meristem, causing bud dormancy and the alteration of the last set of leaves into protective bud covers. Since it was found in freshly abscissed leaves, it was thought to play a role in the processes of natural leaf drop, but further research has disproven this. In plant species from temperate parts of the world, it plays a role in leaf and seed dormancy by inhibiting growth, but, as it is dissipated from seeds or buds, growth begins. In other plants, as ABA levels decrease, growth then commences as gibberellin levels increase. Without ABA, buds and seeds would start to grow during warm periods in winter and be killed when it froze again. Since ABA dissipates slowly from the tissues and its effects take time to be offset by other plant hormones, there is a delay in physiological pathways that provide some protection from premature growth. It accumulates within seeds during fruit maturation, preventing seed germination within the fruit, or seed germination before winter. Abscisic acid's effects are degraded within plant tissues during cold temperatures or by its removal by water washing in out of the tissues, releasing the seeds and buds from dormancy.

In plants under water stress, ABA plays a role in closing the stomata. Soon after plants are water-stressed and the roots are deficient in water, a signal moves up to the leaves, causing the formation of ABA precursors there, which then move to the roots. The roots then release ABA, which is translocated to the foliage through the vascular system and modulates the potassium and sodium uptake within the guard cells, which then lose turgidity, closing the stomata. ABA exists in all parts of the plant and its concentration within any tissue seems to mediate its effects and function as a hormone; its degradation, or more properly catabolism, within the plant affects metabolic reactions and cellular growth and production of other hormones. Plants start life as a seed with high ABA levels. Just before the seed germinates, ABA levels decrease; during germination and early growth of the seedling, ABA levels decrease even more. As plants begin to produce shoots with fully functional leaves, ABA levels begin to increase, slowing down cellular growth in more "mature" areas of the plant. Stress from water or predation affects ABA production and catabolism rates, mediating another cascade of effects that trigger specific responses from targeted cells. Scientists are still piecing together the complex interactions and effects of this and other phytohormones.

Auxins

Auxins are compounds that positively influence cell enlargement, bud formation and root initiation. They also promote the production of other hormones and in conjunction with cytokinins, they control the growth of stems, roots, and fruits, and convert stems into flowers. Auxins were the first class of growth regulators discovered. They affect cell elongation by altering cell wall plasticity. They stimulate cambium, a subtype of meristem cells, to divide and in stems cause secondary xylem to differentiate.

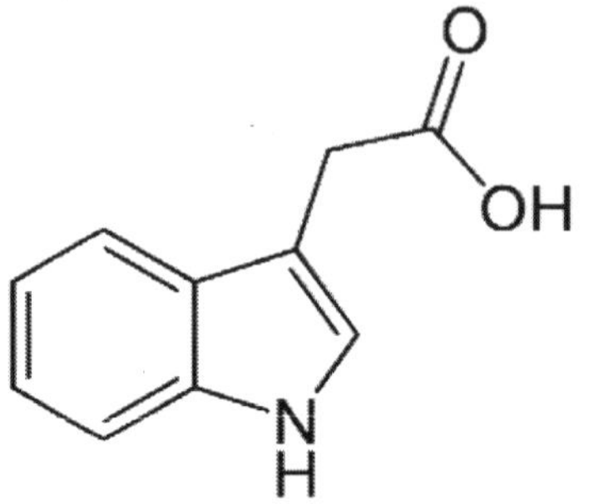

Figure: *The auxin indole-3-acetic acid*

Auxins act to inhibit the growth of buds lower down the stems (apical dominance), and also to promote lateral and adventitious root development and growth. Leaf abscission is initiated by the growing point of a plant ceasing to produce auxins. Auxins in seeds regulate specific protein synthesis, as they develop within the flower after pollination, causing the flower to develop a fruit to contain the developing seeds. Auxins are toxic to plants in large concentrations; they are most toxic to dicots and less so to monocots. Because of this property, synthetic auxin herbicides including 2,4-D and 2,4,5-T have been developed and used for weed control. Auxins, especially 1-Naphthaleneacetic acid (NAA) and Indole-3-butyric acid (IBA), are also commonly applied to stimulate root growth when taking cuttings of plants. The most common auxin found in plants is indole-3-acetic acid or IAA. The correlation of auxins and cytokinins in the plants is a constant (A/C = const.).

Cytokinins

Cytokinins or CKs are a group of chemicals that influence cell division and shoot formation. They were called kinins in the past when the first cytokinins were isolated from yeast cells. They also help delay senescence or the aging of tissues, are responsible for mediating auxin transport throughout the plant, and affect internodal length and leaf growth. They have a highly synergistic effect in concert with auxins, and the ratios of these two groups of plant hormones affect most major growth periods during a plant's lifetime. Cytokinins counter the apical dominance induced by auxins; they in conjunction with ethylene promote abscission of leaves, flower parts, and fruits. The correlation of auxins and cytokinins in the plants is a constant (A/ C = const.).

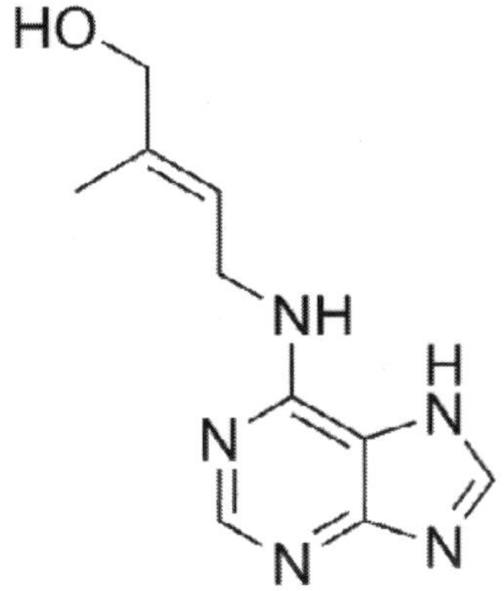

Figure: *The cytokinin zeatin, the name is derived from* Zea, *in which it was first discovered in immature kernels.*

Ethylene

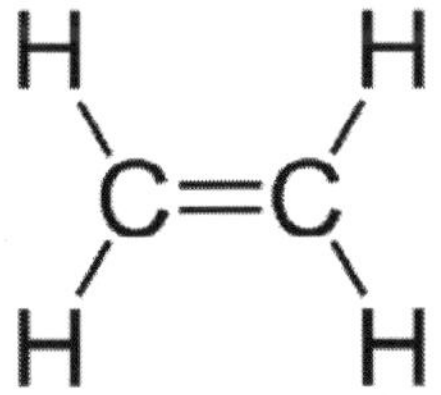

Figure: *Ethylene*

Ethylene is a gas that forms through the breakdown of methionine, which is in all cells. Ethylene has very limited solubility in water and does not accumulate within the cell but diffuses out of the cell and escapes out of the plant. Its effectiveness as a plant hormone is dependent on its rate of production versus its rate of escaping into the atmosphere. Ethylene is produced at a faster rate in rapidly growing and dividing cells, especially in darkness. New growth and newly germinated seedlings produce more ethylene than can escape the plant, which leads to elevated amounts of ethylene, inhibiting leaf expansion.

As the new shoot is exposed to light, reactions by phytochrome in the plant's cells produce a signal for ethylene production to decrease, allowing leaf expansion. Ethylene affects cell growth and cell shape; when a growing shoot hits an obstacle while underground, ethylene production greatly increases, preventing cell elongation and causing the stem to swell. The resulting thicker stem can exert more pressure against the object impeding its path to the surface. If the shoot does not reach the surface and the ethylene stimulus becomes prolonged, it affects the stem's natural geotropic response, which is to grow upright, allowing it to grow around an object. Studies seem to indicate that ethylene affects stem diameter and height: When stems of trees are subjected to wind, causing lateral stress, greater ethylene production occurs, resulting in thicker, more sturdy tree trunks and branches. Ethylene affects fruit-ripening: Normally, when the seeds are mature, ethylene production increases and builds-up within the fruit, resulting in a climacteric event just before seed dispersal. The nuclear protein Ethylene Insensitive2 (EIN2) is regulated by ethylene production, and, in turn, regulates other hormones including ABA and stress hormones.

Gibberellins

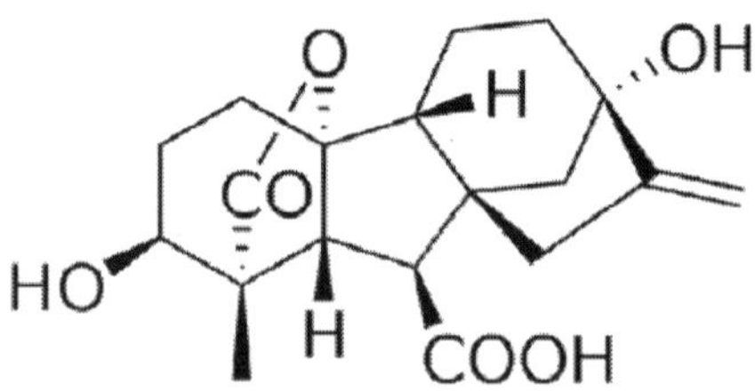

Figure: *Gibberellin A1*

Gibberellins, or GAs, include a large range of chemicals that are produced naturally within plants and by fungi. They were first discovered when Japanese researchers, including Eiichi Kurosawa, noticed a chemical produced by a fungus called *Gibberella fujikuroi* that produced abnormal growth in rice plants. Gibberellins are important in seed germination, affecting enzyme production that mobilizes food production used for growth of new cells. This is done by modulating chromosomal transcription. In grain (rice, wheat, corn, etc.) seeds, a layer of cells called the aleurone layer wraps around the endosperm tissue. Absoption of water by the seed causes production of GA. The GA is transported to the aleurone layer, which responds by producing enzymés that break down stored food reserves within the endosperm, which are utilised by the growing seedling. GAs produce bolting of rosette-forming plants, increasing internodal length. They promote flowering, cellular division, and in seeds growth after germination. Gibberellins also reverse the inhibition of shoot growth and dormancy induced by ABA.

Other Known Hormones

Other identified plant growth regulators include:

- Brassinosteroids, are a class of polyhydroxysteroids, a group of plant growth regulators. Brassinosteroids have been recognised as a sixth class of plant hormones, which stimulate cell elongation and division, gravitropism, resistance to stress, and xylem differentiation. They inhibit root growth and leaf abscission. Brassinolide was the first identified brassinosteroid and was isolated from extracts of rapeseed (*Brassica napus*) pollen in 1979.
- Salicylic acid — activates genes in some plants that produce chemicals that aid in the defence against pathogenic invaders.
- Jasmonates — are produced from fatty acids and seem to promote the production of defence proteins that are used to fend off invading organisms. They are believed to also have a role in seed germination, and affect the storage of protein in seeds, and seem to affect root growth.
- Plant peptide hormones — encompasses all small secreted peptides that are involved in cell-to-cell signaling. These small peptide hormones play crucial roles in plant growth and development, including defence mechanisms, the control of cell division and expansion, and pollen self-incompatibility.
- Polyamines — are strongly basic molecules with low molecular weight that have been found in all organisms studied thus far. They are essential for plant growth and development and affect the process of mitosis and meiosis.
- Nitric oxide (NO) — serves as signal in hormonal and defence responses (e.g. stomatal closure, root development, germination, nitrogen fixation, cell death, stress response). NO can be produced by a yet undefined NO synthase, a special type of nitrite reductase, nitrate reductase, mitochondrial cytochrome c oxidase or non enzymatic processes and regulate plant cell organelle functions (e.g. ATP synthesis in chloroplasts and mitochondria).
- Strigolactones, implicated in the inhibition of shoot branching.
- Karrikins, a group of plant growth regulators found in the smoke of burning plant material that have the ability to stimulate the germination of seeds.

Potential Medical Applications

Plant stress hormones activate cellular responses, including cell death, to diverse stress situations in plants. Researchers have found that some plant stress hormones share the ability to adversely affect human cancer cells. For example, sodium salicylate has been found to suppress proliferation of lymphoblastic leukemia, prostate, breast, and melanoma human cancer cells. Jasmonic acid, a plant stress hormone that belongs to the jasmonate family, induced death in lymphoblastic leukemia cells. Methyl jasmonate has been found to induce cell death in a number of cancer cell lines.

Hormones and Plant Propagation

Synthetic plant hormones or PGRs are commonly used in a number of different techniques involving plant propagation from cuttings, grafting, micropropagation, and tissue culture.

The propagation of plants by cuttings of fully developed leaves, stems, or roots is performed by gardeners utilising auxin as a rooting compound applied to the cut surface; the auxins are taken into the plant and promote root initiation. In grafting, auxin promotes callus tissue formation, which joins the surfaces of the graft together. In micropropagation, different PGRs are used to promote multiplication and then rooting of new plantlets. In the tissue-culturing of plant cells, PGRs are used to produce callus growth, multiplication, and rooting.

Seed Dormancy

Plant hormones affect seed germination and dormancy by acting on different parts of the seed.

Embryo dormancy is characterized by a high ABA:GA ratio, whereas the seed has a high ABA sensitivity and low GA sensitivity. In order to release the seed from this type of dormancy and initiate seed germination, an alteration in hormone biosynthesis and degradation toward a low ABA/GA ratio, along with a decrease in ABA sensitivity and an increase in GA sensitivity, must occur.

ABA controls embryo dormancy, and GA embryo germination. Seed coat dormancy involves the mechanical restriction of the seed coat. This, along with a low embryo growth potential, effectively produces seed dormancy. GA releases this dormancy by increasing the embryo growth potential, and/or weakening the seed coat so the radical of the seedling can break through the seed coat.

Different types of seed coats can be made up of living or dead cells, and both types can be influenced by hormones; those composed of living cells are acted upon after seed formation, whereas the seed coats composed of dead cells can be influenced by hormones during the formation of the seed coat. ABA affects testa or seed coat growth characteristics, including thickness, and effects the GA-mediated embryo growth potential. These conditions and effects occur during the formation of the seed, often in response to environmental conditions. Hormones also mediate endosperm dormancy: Endosperm in most seeds is composed of living tissue that can actively respond to hormones generated by the embryo. The endosperm often acts as a barrier to seed germination, playing a part in seed coat dormancy or in the germination process. Living cells respond to and also affect the ABA/GA ratio, and mediate cellular sensitivity; GA thus increases the embryo growth potential and can promote endosperm weakening. GA also affects both ABA-independent and ABA-inhibiting processes within the endosperm.

Plant Growth Promoters

The growth of plants is regulated by certain chemical substances, which are synthesized by the plant in very small quantities. These substances are formed in one tissue or organ of the plant and are then transported to other sites where they produce specific effects on growth and development. They are referred to as plant hormones. Plant hormones are organic compounds which are capable of promotion, inhibition or modification of growth. The plant hormones are also known as growth factors, growth hormones, growth substances, growth regulators or phytohormones.

Phytohormones are grouped into 2 main types - growth promoters have a positive effect on a process and thus promote it, whereas the growth inhibitors have a negative effect and cause inhibition. A particular hormone may promote certain processes, inhibit some others and not affect many others. They act synergistically (co-operative and beneficial) or antagonistically (acting in opposition) with one another.

Auxins

Auxins (in Greek - auxein means to grow) are one of the most important groups of plant hormones because of their many sided role in plants. These substances were also the first growth factors identified as plant hormones. The principal naturally occurring auxin is indole-3 acetic acid - IAA.

History

In the last sixty years, a large number of growth regulators have been isolated from plants.

The first indication of their existence was given by Darwin (1880). Working on canary grass (Phalaris) he found that if a unilateral source of light was given, the coleoptile would bend towards the source of light. He believed that the tip contained a substance which was transmitted to the lower portion where it caused a curve. He also demonstrated that a decapitated coleoptile or a coleoptile which was covered with a tin foil cap failed to respond to unilateral light. But curvature was seen when seedling was buried in fine black sand with tip exposed.

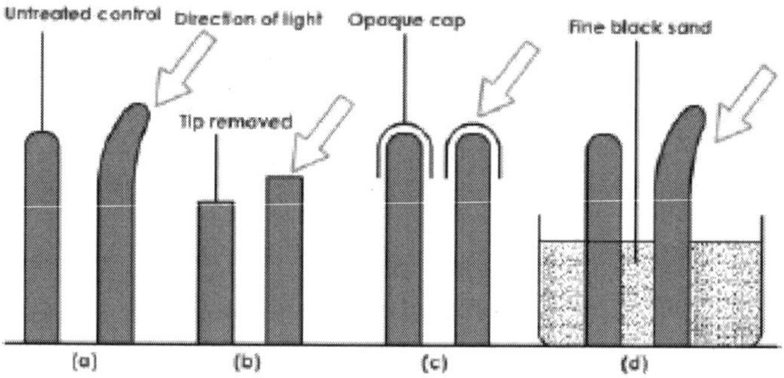

Figure: *Darwin's Experiment with the Seeding of Canary Grass*

Boysen Jensen (1910 - 1913) of Denmark found that the phototropic response lost by decapitation of the tip could be recovered if the tip was replaced on the stump. He further demonstrated that if a transverse slit was caused in the coleoptile on the dark side and a piece of mica inserted into the slit, no phototropic response took place. On the other hand there was a phototropic response if the slit and the piece of mica were on the illuminated side. He concluded that a substance migrates down the dark side promoting growth curvature towards light.

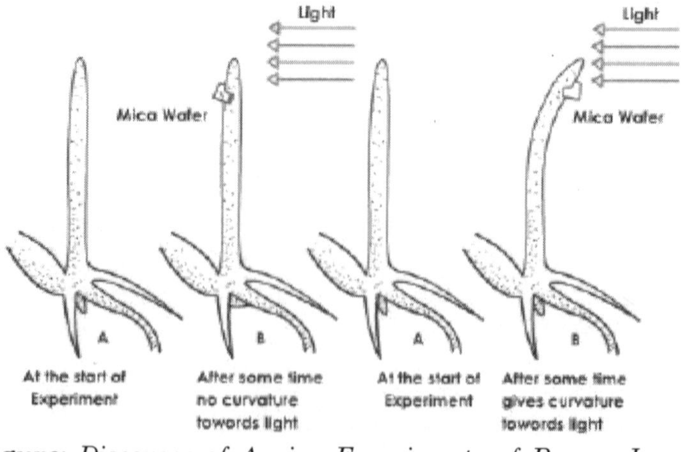

Figure: *Discovery of Auxins Experiments of Boysen-Jensen*

Went (1928) from Holland demonstrated the existence of a substance in the tip of Oat (Avena) seedling which diffused downwards and promoted growth. He demonstrated that when the tip of Oat coleoptile was removed, growth stopped. When the freshly cut coleoptile tip was placed on an agar block, for a few hours so as to allow the growth substances to diffuse into, and the agar block was then placed on the cut end of the coleoptile, growth occurred.

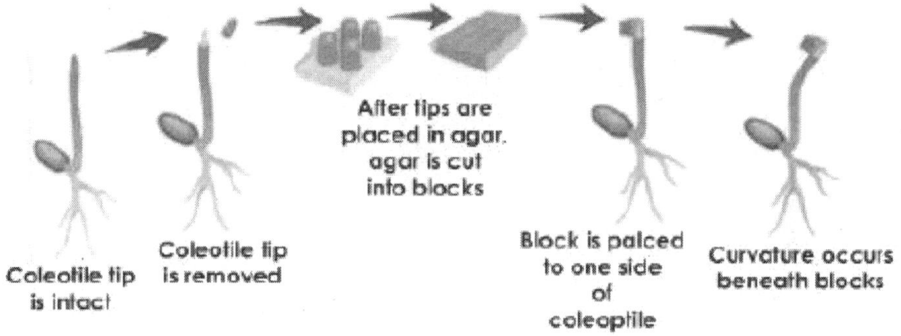

Figure: *Oat Coleoptile Experiment*

If the agar block was placed laterally on the cut end of the coleoptile, only that side of the coleoptile elongated resulting in a curvature. The side of the coleoptile that received the growth substance elongated faster and caused the curvature towards the opposite side. He called this substance Auxin from the Greek word 'auxein' to grow.

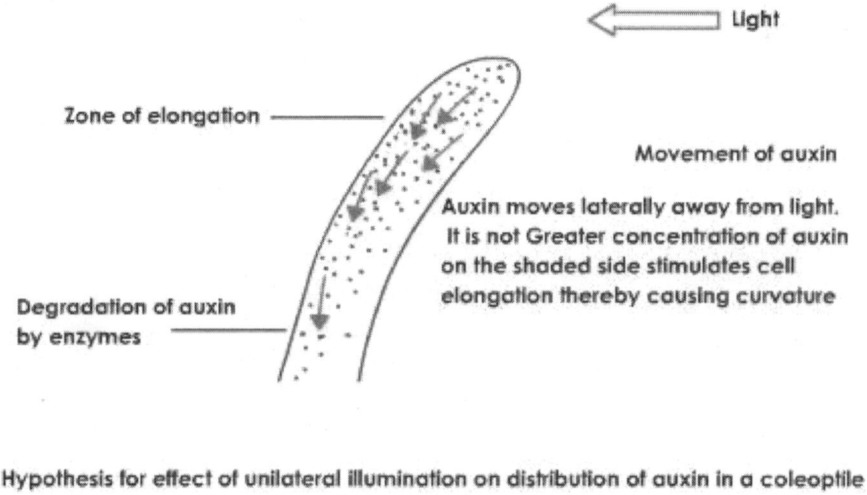

Figure: *Hypothesis for Effect of Unilateral Illumination on Distribution of Auxin in a Coleopite*

It is interesting to note that the first auxin was isolated from human urine of persons suffering from pellagra.

Functions of Auxins: Auxins control several plant growth processes.

Cell Elongation: Auxins promote elongation and growth of stems and roots and enlargement of many fruits by stimulating elongation of cells in all directions.

Reactivation of Cambium: Auxins promote cell division in vascular cambium. The reactivation of cambium in the growing season is due to the moving of IAA from the developing shoot buds.

Apical Dominance: Auxins induce apical dominance, where the apical bud suppressed the growth of lateral buds.

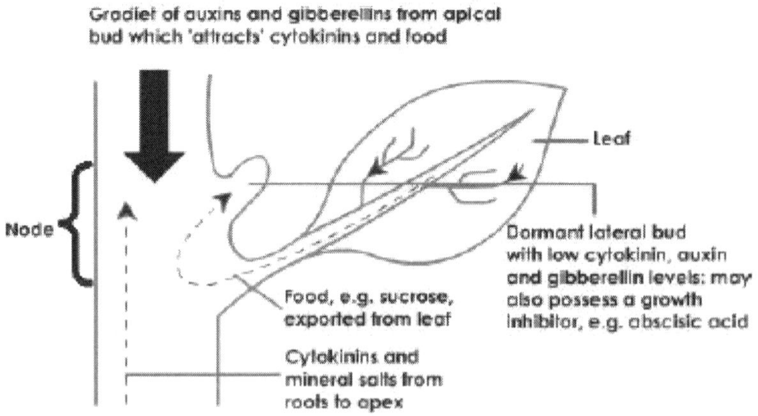

Figure: *Possible Involvement of Plant Growth Substances in Apical Dominance in Presence of Apical Bud*

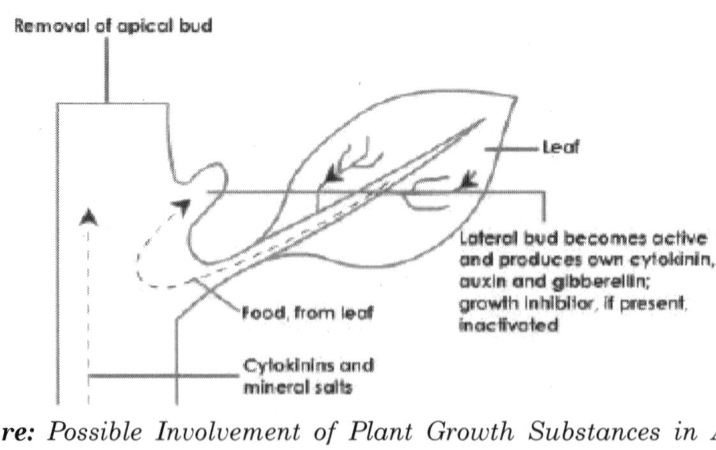

Figure: *Possible Involvement of Plant Growth Substances in Apical Dominance after Removal of Apical Bud*

Inhibition of Abscission

Formation of an abscission layer at the base of petiole or pedicel results in shedding of leaves, flowers or fruits. But auxins inhibit abscission, as they prevent the formation of abscission layer.

a) b)

Figure: *Auxin Spray Prevents Premature Fruit Abscission and Increase in Yield.a) Auxin Sprayed; b) Auxin not Sprayed*

Parthenocarpy

Auxin induces parthenocarpy, i.e., the formation of seedless fruits without the act of fertilisation.

Development of Seedless Fruits

Root Formation: IAA stimulates cell division in the pericylce leading to the formation of lateral and adventitious roots. Auxin stimulates formation of new roots, but inhibits the root growth.

As Inhibitor

Higher concentration of auxins inhibit the growth and exert toxic effect on plants. In nature auxin inhibits the growth of lateral buds.

Uses of Auxins in Agriculture

Auxins of various types are used now-a days for certain useful aspects of agricultural practices directly related to economic gains and food production.

Weedicides

Many synthetic auxins are used as selective weed killers and herbicides. 2, 4 - D (2, 4 - dichloro phenoxy acetic acid) is used to destroy broad leaved weeds. It does not affect mature monocotyledonous plants.

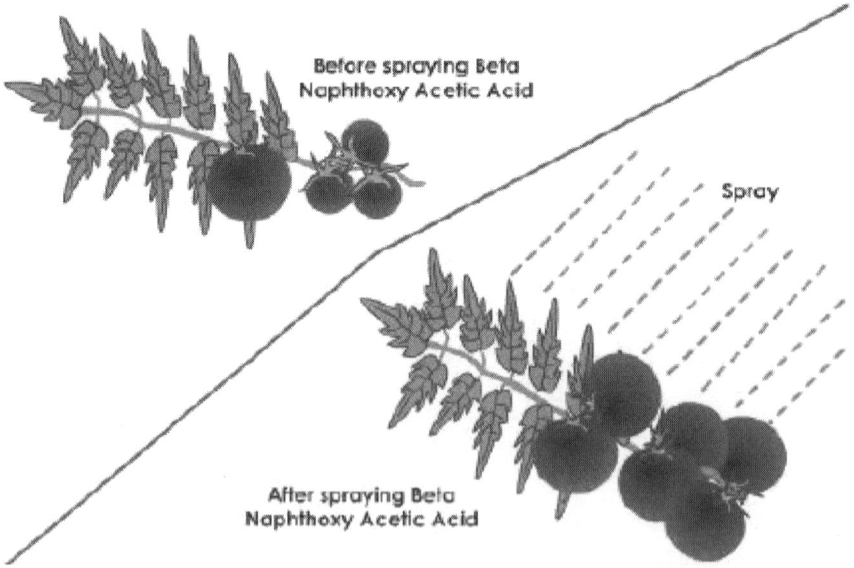

Figure: *Destruction of Weeds by 2,4-D Spray*

Rooting

Naphthalene acetic acid (NAA), indole butyric acid (IBA) and indole acetic acid (IAA) are used to induce rooting of cuttings of woody plants like guava.

Flowering

Foliar spray of NAA and 2, 4 - D induces flowering in litchi and pineapple. Percentage of ball setting in cotton plants increases when NAA or IBA are applied.

Pre-Mature Fruit Drop

Auxins such as 2, 4 - D, IAA, IBA have been used successfully to prevent premature fruit drop in apples, pears and oranges.

Parthenocarpy

NAA and IBA treatment induces parthenocarpy of fruits in tomatoes, ladies finger and brinjal.

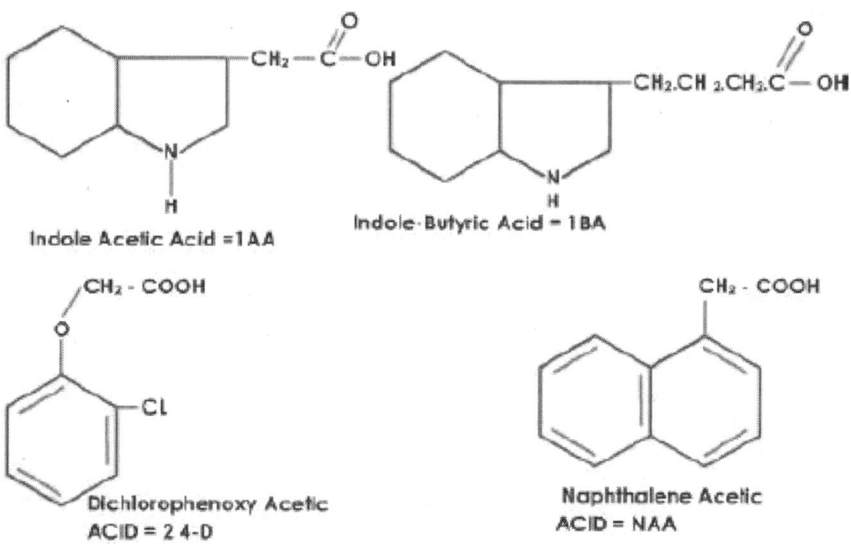

Figure: *Some Common Auxins*

Dormancy: Dormancy of seeds can be broken by auxin application.

Storage: The methyl ester of NAA is used to prevent the sprouting of potato tubers.

Vegetable Crops: Auxins improve the quality of vegetable crops by inhibiting flower formation in some plants.

Fruits: Auxins such as IBA, increase the sweetening of fruits in many plants.

Misuse: Synthetic auxins have been misused by people. Large scale application of defoliants used in Vietnam to expose the forests, exterminated the wild relatives of economically useful plants such as Citrus.

Auxins are of 2 Types

Natural Auxins: These are naturally occurring in plants. The best known and universally present natural auxin is Indole - 3 - acetic acid. Other natural auxins are Indole - 3 - pyuruvic acid, Indole - 3 - ethanol, Indole - 3 - acetaldehyde.

Synthetic Auxins

These are the chemicals synthesized by chemists that cause various physiological actions similar to IAA. Some of the synthetic auxins are Indole - 3 - butyric acid (IBA), b - napthalene acetic acid (NAA) and 2,4 - dichlorophenoxy acetic acid (2,4 - D).

Auxin precursors - They are compounds which can be converted into auxins.

Antiauxins

These are chemicals which inhibit the action of auxins.

Examples: 2, 3, 5 triiodobenzoic acid (TIBA) and napthylthalmic acid (NPA).

Free auxins - are auxins which can be easily extracted. They are active.

Bound auxins - are auxins which are difficult to extract. They are inactive.

Training and Pruning

Training, the orienting of the plant in space, is achieved by techniques that direct the shape, size, and direction of plant growth. It may be accomplished by use of supports to which plants can be bent, twisted, or fastened. Pruning, the judicious cutting away of plant parts, is performed for other purposes: to contain size, to encourage fruiting in orchard trees, or to improve the appearance of ornamental trees and shrubs. It is one of the most important horticultural arts.

Where trees and shrubs are left to grow naturally, they often become much too large for their space in the garden. Also they may grow lanky and misshapen and have much dead growth. Where a branch or shoot is cut, it will often be induced to make a number of young shoots from below the cut, and these are likely to flower more freely than the older branches. Fruit trees in particular when pruned annually often give fruit of finer quality, larger in size, freer of disease, and of better colour. The two basic pruning cuts are known as heading back and thinning out. Heading back consists of cutting back the terminal portion of a branch to a bud; thinning out is the complete removal of a branch to a lateral or main trunk. Heading back, usually followed by the stimulation of lateral budbreak below the cut, produces a bushy, compact plant, suitable for a hedgerow, and it is often used to rejuvenate shrubs that have become too large or that flower poorly.

Thinning out, which encourages longer growth of the remaining terminals by reducing lateral branches, tends to open up the plant, producing a longer plant. In general, pruning, started when the plant is young, obviates the need for drastic and risky remedial pruning later of a large, old, or misshapen bush or tree. Particular spatial arrangements may increase light utilisation, facilitate harvesting or disease control, or improve productivity and quality. Thus, training and pruning form an essential part of fruit growing throughout the life of the plant. Special attention is given in the formative years to obtain desired shape and structure.

The key to training is the point on the main stem from which branches form. In the central-leader system of training, the trunk forms a central axis with branches distributed laterally up and down and around the stem. In the open-centre or vase system, the main stem is terminated and growth forced through a number of branches originating close to the upper end of the trunk. An intermediate system is called the modified-leader system. In espalier systems plants are trained to grow flat along a wire or trellis. Properly executed espaliers are extremely attractive as ornamentals. Espaliers in combination with dwarfing rootstocks allow high-density orchards that are very productive on a per-unit-area basis, with the fruit close to the ground for easy harvest. Extensive pruning is required annually to maintain the system.

There are a number of physiological responses to training and pruning. Orientation of the plant may have a marked effect on growth and fruiting. Thus fruit trees planted on an inclined angle become dwarfed and flower earlier; training branches in a horizontal position produces the same effect. This effect is achieved naturally when a heavy fruit load bends a limb down. The main effects of pruning are achieved by altering the root–shoot balance. Thus an explosion of vegetative growth normally occurs after extensive shoot pruning. Severely pruned plants, especially if they are in the juvenile stage of growth, tend to remain vegetative. Similarly the slowdown of vegetative growth by root pruning encourages flowering.

Pruning

Pruning is a horticultural and silvicultural practice involving the *selective* removal of parts of a plant, such as branches, buds, or roots. Reasons to prune plants include deadwood removal, shaping (by controlling or directing growth), improving or maintaining health, reducing risk from falling branches, preparing nursery specimens for transplanting, and both harvesting and increasing the yield or quality of flowers and fruits.

The practice entails *targeted* removal of diseased, damaged, dead, non-productive, structurally unsound, or otherwise unwanted tissue from crop and landscape plants. Specialised pruning practices may be applied to certain plants, such as roses, fruit trees, and grapevines. Different pruning techniques may be deployed on herbaceous plants than those used on perennial woody plants. Hedges, by design, are usually (but not exclusively) maintained by hedge trimming, rather than by pruning. Arborists, orchardists, and gardeners use various garden tools and tree cutting tools designed for the purpose, such as hand pruners, loppers, or chainsaws. In nature, meteorological conditions such as wind, ice and snow, and salinity can cause plants to self-prune.

Pruning Landscape and Amenity Trees

Types of Branch Union

Figure: *Pruning when there's a branch collar: Note the swollen area where the branch joins the trunk, this is known as the collar, do not cut off the collar*

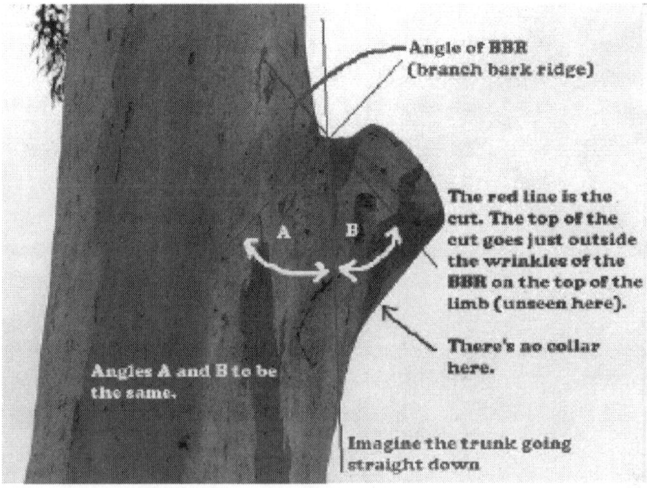

Figure: *Pruning when there's a collarless union: Note there's no swollen area where the branch joins the trunk, this is known as a collarless union, cut at a mirror angle to the BBR (branch bark ridge)*

Figure: *Pruning when it's a codominant stem: Note the bottom of the BBR and where it meets the outside of the stem, cut where the red line is.*

For arboricultural purposes the unions of tree branches (i.e. where they join together) are placed in one of three types: collared, collarless or codominant. Regardless of the overall type of pruning being carried out, each type of union is cut in a particular way so that the branch has less chance of regrowth from the cut area and best chance of sealing over and compartmentalising decay. This is often referred to by arborists as "target cutting".

Dead-wooding

Branches die off for a number of reasons ranging from light deficiency, pests and disease damage to root damage. A dead branch will at some point decay back to the parent stem causing abscission and fall off. This is normally a slow process but can be shortened by high winds and extremities of temperature. Therefore the main reason deadwooding is carried out is safety. The situations that usually demand such removal would normally be trees that overhang public roads, houses, public areas and gardens. Trees that are located in wooded areas are usually considered and assessed as lower risk but such assessments would need to consider the amount of visitors. Usually, trees adjacent to the footpaths and access roads are considered for deadwood removal. Another reason for deadwooding is amenity value, i.e. a tree with a large amount of dead throughout the crown looks more aesthetically pleasing with the deadwood removed. The physical practice of deadwooding can be carried out most of the year although preferably not when the tree is coming into leaf. The deadwooding process will speed up the natural abscission process the tree follows. It will help remove unwanted weight; wind resistance the tree carries and can help the overall balance.

Crown and Canopy Thinning

Increase light and reduce wind resistance by selective removal of branches throughout the canopy of the tree. This is a common practice which improves the tree's strength against adverse weather conditions as the wind can pass through the tree resulting in less "load" being placed on the tree.

Crown Canopy Lifting

Crown lifting involves the removal of the lower branches to a given height. The height is achieved by the removal of whole branches or removing the parts of branches which extend below the desired height. The branches are normally not lifted to more than one third of the tree's total height.

Crown lifting is done for access; these being pedestrian, vehicle or space for buildings and street furniture. Lifting the crown will allow traffic and pedestrians to pass underneath safely. This pruning technique is usually used in the urban environment as it is for public safety and aesthetics rather than tree form and timber value.

Crown lifting introduces light to the lower part of the trunk; this, in some species can encourage epicormic growth from dormant buds. To reduce this sometimes smaller branches are left on the lower part of the trunk. Excessive removal of the lower branches can displace the canopy weight, this will make the tree top heavy, therefore adding stress to the tree. When a branch is removed from the trunk, it creates a large wound. This wound is susceptible to disease and decay, and could lead to reduced trunk stability. Therefore much time and consideration must be taken when choosing the height the crown is to be lifted to.

This would be an inappropriate operation if the tree species' form was of a shrubby nature. This would therefore remove most of the foliage and would also largely unbalance the tree. This procedure should not be carried out if the tree is in decline, poor health or dead, dying or dangerous (DDD) as the operation will remove some of the photosynthetic area the tree uses. This will increase the decline rate of the tree and could lead to death.

If the tree is of great importance to an area or town, (i.e. veteran or ancient) then an alternative solution to crown lifting would be to move the target or object so it is not in range. For example, diverting a footpath around a tree's drip line so the crown lift is not needed. Another solution would be to prop up or cable-brace the low hanging branch. This is a non-invasive solution which in some situations can work out more economically and environmentally friendly.

Directional or Formative Pruning

Removal of appropriate branches to make the tree structurally sound whilst shaping it.

Vista Pruning

Selectively pruning a window of view in a tree.

Crown Reduction

Reducing the height and or spread of a tree by selectively cutting back to smaller branches And in fruit trees for increasing of light interception and enhancing fruit quality.

Pollarding

A regular form of pruning where certain deciduous species are pruned back to pollard heads every year in the dormant period. This practice is usually commenced on juvenile trees so they can adapt to the harshness of the practice.

Types of Pruning

Regardless of the various names used for types of pruning, there are only two basic cuts: One cuts back to an intermediate point, called *heading back cut*, and the other cuts back to some point of origin, called *thinning out cut.*

Removing a portion of a growing stem down to a set of desirable buds or side-branching stems. This is commonly performed in well trained plants for a variety of reasons, for example to stimulate growth of flowers, fruit or branches, as a preventative measure to wind and snow damage on long stems and branches, and finally to encourage growth of the stems in a desirable direction. Also commonly known as *heading-back.*

- Thinning: A more drastic form of pruning, a thinning out cut is the removal of an entire shoot, limb, or branch at its point of origin. This is usually employed to revitalize a plant by removing over-mature, weak, problematic, and excessive growths. When performed correctly, thinning encourages the formation of new growth that will more readily bear fruit and flowers. This is a common technique in pruning roses and for implifying and "opening-up" the branching of neglected trees, or for renewing shrubs with multiple branches.
- Topping: Topping is a very severe form of pruning which involves removing all branches and growths down to a few large branches or to the trunk of the tree. When performed correctly it is used on very young trees, and can be used to begin training younger trees for pollarding or for trellising to form an espalier.
- Raising removes the lower branches from a tree in order to provide clearance for buildings, vehicles, pedestrians, and vistas.
- Reduction reduces the size of a tree, often for clearance for utility lines. Reducing the height or spread of a tree is best accomplished by pruning back the leaders and branch terminals to lateral branches that are large enough to assume the terminal roles (at least one-third the diameter of the cut stem). Compared to topping, reduction helps maintain the form and structural integrity of the tree.

In orchards, fruit trees are often lopped to encourage regrowth and to maintain a smaller tree for ease of picking fruit. The pruning regime in orchards is more planned and the productivity of each tree is an important factor.

Deadheading is the act of removing spent flowers or flowerheads for aesthetics, to prolong bloom for up to several weeks or promote rebloom, or to prevent seeding.

Technique

The general rule of pruning is to always cut in a location where growth will occur, whether the cut is next to a bud or another branch. Cutting a branch beyond where growth will occur will prevent the plant from forming a callus over the cut surface, which in turn will invite insects and infection. It effectively kills all portions of that branch back to the closest branch, bud, or dormant bud clusters, leaving a stub of dead wood. The withered stub will eventually rot away and fall off. All cuts should be relatively smooth since this will aid in healing. Also, the pruning cut should not be too large when compared to the growing point. For instance, a large cut on a 20 cm trunk down to a 15 cm branch should be fine, but the same cut to the trunk down to a 1 cm twig or bud is considerably less ideal and should be avoided if possible.

Pruning to a Bud

A correct pruning cut will allow for quick healing and promote vigorous growth from the closest bud to the cut. The cut should be close enough to the bud to reduce the size of the stub of dead wood that will form from the cut, but far enough away to prevent the bud from being adversely affected by the cut through desiccation. Cutting too close to the bud (under-cutting) sometimes results in the death of the bud, which results in a scenario similar to cutting too far away from the bud (over-cutting). In general, a correct cut should be angled at a moderate 35-45 degree slant such that its lowest point is situated on the same level as the tip of the growth bud. This technique is usually applied when pinching or when cutting-back.

Pruning to a Main Branch

The pruning cut should occur slightly away from and follow the branch collar. When cutting away branches growing directly from the roots, the cut should be flush and level to the ground. This technique is usually applied when thinning or to remove larger dead or damaged branches.

When using pruning shears or loppers to remove a branch back to a main branch, the "hook" portion of the shears should always face away from the main branch. This ensures that the blade will not leave a protruding stub and the hook will not damage the branch collar or parts of the main branch.

Large Heavy Branches

Depending on the weight of the branch, the first cut should be a notch on the underside of the branch about a third to half of the way through. The bulk of the branch should then be removed with a follow-through cut slightly above the first cut, thus leaving a limb stub. The purpose of this is to stop the weight of the branch from tearing the bark of the tree from the underside, which would normally occur if the removal was done with one cut. The limb stub ensures that any cracking of the wood resulting from the branch separation is limited to the portion of the wood to be removed. The branch collar should then be located, and can be identified by the strip of rough bark running down from the topside of the branch at its junction with the stem. The cut for removing the limb stub should be just outside the branch collar, leaving a small bump. The bump and the branch collar should not be removed since this action can increase healing time, which could result in a major infection.

Time Period

Pruning small branches can be done at any time of year. Large branches, with more than 5-10% of the plant's crown, can be pruned either during dormancy in winter, or, for species where winter frost can harm a recently-pruned plant, in mid summer just after flowering. Autumn should be avoided, as the spores of disease and decay fungi are abundant at this time of year.

Some woody plants that tend to bleed profusely from cuts, such as maples, or which callous over slowly, such as magnolias, are better pruned in summer or at the onset of dormancy instead. Woody plants that flower early in the season, on spurs that form on wood that has matured the year before, such as apples, should be pruned right after flowering, as later pruning will sacrifice flowers the following season. *Forsythia*, azaleas and lilacs all fall into this category.

How to Trim Vegetable Plants for Maximum Growth

Trimming or pruning vegetables can help divert nutritional energy for improved growth. Branching stems and growth that does not contain flower buds can be safely cut away so the plant produces more vegetables and flowers. The correct time to do this depends on the variety of flowers and vegetables but is typically just before blooms develop.

Instructions

1. Take stock of the plants in the garden, when each variety was planted, and the rate at which it develops. This helps you plan which plants to trim and prune first. For example, ground and vine plants such as cucumbers, squash and zucchini will be

ready to prune before tomatoes, if all were planted at the same time.

2. Cut non-flowering leaves and stems in an alternating pattern on both sides of the plant. For example, clip leaves at the mid-section on the left side, then clip higher or lower on the right side, leaving buds intact.
3. Prune dead flower blooms and stems of harvested vegetables to route nutrients toward areas of the plant that are still productive.
4. Trim dead or dying areas of the plant so the hearty areas continue to thrive.

Water Management

Water is a vital component of vegetable crop production. An adequate water supply is essential to maximise both quality and crop yield for a given date. It is also an important management tool, enhancing crop establishment, chemical weed control and harvesting of root vegetables. However, water has to be applied at the correct time, in sufficient quantities, but without waste and with sympathy for the environment as a whole. We know that water supplies are frequently stretched and action has to be taken to safeguard this essential resource. Legislation expected to come into force this year will change the basis of licensing water abstraction. It will for the first time include water used for trickle irrigation.

It is probable that the demand for, as well as the cost, of water will increase. This will mean that all users will need to increase water and irrigation management efficiency to make the most of this basic resource. Environmental considerations are increasingly playing a part in the growing and marketing of crops. Careful and effective crop monitoring and irrigation control will be critical to profitable cropping in the future.

Water Quality

The quality of water to be used for irrigation is important. All water is not the same. Mains and borehole water sources offer more consistency than surface water. They are unlikely to have the potential problems that can be caused by biological or herbicide contamination, by soil particles or organic matter. The two main aspects of water quality are its chemical and microbiological properties. Other factors can occasionally affect quality, especially from river or reservoir sources, such as the presence of suspended peat or silt particles or other specific pollutants (e.g. herbicides). The chemical parameters important for irrigation water include:

- pH
- Alkalinity or bicarbonate content
- Electrical conductivity (measure of the total salts content)
- Chloride level
- Sulphate level
- Iron and other trace element levels

(Note that some trace elements such as iron and manganese, although often present in water, are not in a form that can be readily taken up by plants.) pH and Alkalinity pH is the measure of how acid or alkaline water is.

However, it is the level of bicarbonates (principally calcium, but also magnesium and sodium) in the water that determines whether it is termed either 'hard' or 'soft'. Water may, for example, have a high pH but a low calcium bicarbonate content.

Whilst it is not practical or economic to consider acidification for field grown crops, it may be worth considering the use of acidifying fertiliser mixes when fertigating with hard water in trickle systems used on high value vegetable crops.

Electrical Conductivity and Trace Elements

The electrical conductivity (E.C. or sometimes referred to as cF) is a measure of the total salts content of the water, reported in micro-siemens per cm. (ìS/cm). The main contributors to high conductivity levels are nitrate, chloride and sulphate ions. Water with a naturally high E.C. does not usually give rise to a build up of salts in soils used for field vegetable production due to the diluting effect of winter rainfall. Rainwater is generally uncontaminated, it has a low conductivity and pH and a very low calcium level. However, borehole water in coastal and fenland areas may have a high conductivity due to salt water contamination. A high conductivity due to high chloride levels makes the water unsuitable for irrigation unless treated or mixed with water from a more suitable source.

Crops Do Vary in Their Susceptibility to Chloride Levels

Some plants are also known to be sensitive to high fluoride levels; therefore irrigation water with less than 0.25 ppm fluoride is desirable. Boron toxicity is occasionally found where water (e.g. river water) contains high boron levels originating from detergents.

Microbiological Contamination

The main microbiological contaminants of water are fungal spores (e.g. white rot, Sclerotinia of Alliums). Irrigation can also provide an entry point for bacteria, such as E. coli, into the food chain. This is of particular significance where water is applied to crops eaten raw, such as salads.

A risk assessment of water source and period when contamination is most likely to occur should be undertaken. Irrigation water should be checked to ensure it has an acceptably low level of microbial contaminants before use, though sampling where abstraction is direct from river and stream water is less representative because of fluctuating bacterial loading over time.

Water obtained from deep boreholes is likely to contain fewer pathogens; water extracted from rivers and stored in reservoirs prior to application will show a decline in pathogen levels over time. Drip irrigation, because it avoids wetting the aerial parts of the plant, may result in reduced bacterial loadings. However, there is little, if any, information to date on pathogen incidence in irrigation waters at point of use.

Water Analysis

Hand-held equipment is available for monitoring E.C. and pH during the production season, but a full laboratory water analysis is recommended at least every 12 months for each water source used and for surface water abstraction at times of high risk (e.g. dry periods).

Seasonal Water Requirements and Storage

The feasibility of water storage will relate to water availability and source. These can be summarised as:

- Summer surface water directly abstracted for irrigation
- Winter surface water for reservoirs
- Boreholes, including wells and well points – direct supply or for reservoir storage
- Public mains supply

The move to a Catchment Abstraction Management Strategy, designed to balance the needs of water users and the environment, will enable the Environment.

Agency to operate a transparent system where availability of further water supplies is evident. Currently, it is unlikely that new or increased abstraction licences will be granted for summer surface water abstraction. Similarly, there are many areas where new licences for borehole abstraction are unlikely to be granted.

Other sources require storage facilities. For winter abstraction of surface water, a reservoir is required on farm to store the predicted annual requirement. Mains water systems are more appropriate for smaller scale production areas and plant raising nurseries. Here sufficient water should be stored to meet one or two day's use at maximum demand.

Growers need to produce a high-quality product efficiently to remain competitive, however, soil and water resources must also be preserved. Healthy, productive plants require healthy soil and clean water.

Soil and Water Management

Intensive vegetable production, whether for processing or fresh market, returns little organic matter to the soil. Tillage used to prepare the seedbed increases the loss of organic matter. To maintain or increase organic matter levels:

- Use cover crops within the rotation. Following short-season vegetables, establish a cover crop as soon as possible. This green manure crop increases organic matter levels and also breaks some pest life cycles.
- For long-season vegetables, annual or cereal rye is usually the best cover crop. It grows well in cooler weather such as late autumn and early spring. Rye's large, fibrous roots help hold the soil together, preventing erosion. Tillage or herbicides will kill the rye prior to spring planting.
- When a cereal crop precedes vegetables, underseed the cereal with either clover or alfalfa to improve soil structure and reduce compaction. Legumes produce nitrogen, so make allowances in your nutrient applications.
- Reduce tillage and add manure, mushroom compost (a permit is required by Ministry of the Environment for the application of organic off-farm waste) and other plant waste. Take care not to increase soil compaction. Adjust the following year's nutrient application depending on the content of these materials.

Cereal crops like wheat make a good break in a crop rotation, helping to build and maintain soil organic matter and soil structure.

Soil Compaction

Soil compaction is a growing concern for vegetable producers. Increased mechanization has led to larger and heavier equipment to ensure planting and harvesting are handled on time. Seedbed preparation and harvest operations under wet soil conditions are the major causes of soil compaction. Perishability and maturity of the vegetable crop are important to quality. Because staying out of wet fields is often not an option, research continues into solutions.

Crop Rotation

Crop rotation is a best management practice for vegetable growers. It will address loss of organic matter, disease, weed and insect pressures, soil nutrition, compaction and erosion. Two rules of thumb:

- The longer the rotation, the better.
- Rotate between different families of crops.

In designing a rotation, ask yourself the following questions:

- Is the rotation profitable?
- Are the yields sustainable?
- Does the sequence allow the use of cover crops?
- Does it make use of nitrogen produced by an earlier crop?
- Does it allow for timely planting and harvest?
- Are harmful herbicide residues left?

Recent tomato research shows that yields improve with good crop rotations. Building and maintaining soil resources should produce similar results for all vegetable crops.

Processing peas can be particularly hard on soil structure. Tightly scheduled planting and harvest seasons mean soil moisture levels may not be optimum when machinery, such as pea combines, are running over the soil. Early or short season crops such as melons allow the use of cover crops and green manure crops to build and maintain soil organic matter.

Wind and Water Erosion

Level sandy soils are at the highest risk of wind erosion while hilly fields are also subject to water erosion. Windbreaks, grassed waterways and other structures address problems in the long-term. Increased residue on the soil surface and use of cover crops will help in the short-term.

For precision-seeded crops, choose a field sheltered by a windbreak, woodlot or other means. Strip cropping with another crop will also cut down wind. The cereal will shelter the seedlings. A timely application of contact grass herbicide will kill the cover crop before it competes with the vegetable.

In some transplanted vegetable production systems, ground beds roughen the soil surface which slows water, wind and soil movement. Some growers are also managing cover crops on ground beds to control wind. Another alternative is the use of narrow grass strips spaced across a field to reduce the speed and soil-carrying ability of wind.

A rye cover crop on beds can be managed with a timed application of herbicides to provide short-term wind protection in the spring. This system also minimizes the number of passes over the field in spring.

Irrigation

Average rainfall is irregular and sometimes is inadequate for vegetables. Irrigation can be profitable with high-value vegetable crops. Both overhead and sprinkler irrigation systems are being used in Ontario. Overall costs are comparable. Drip irrigation has two advantages: reduced water needs and uniform soil moisture; but, overhead irrigation is adaptable to any crop and can be used for frost protection.

Irrigation is important after planting until seedlings emerge and during fruit development. Most vegetables have periods where a lack of water can affect yield and quality. Use a scheduling method such as the tensiometre or the evapotranspiration model to assist in irrigation timing.

Plasticulture

This practice combines plastic mulches with row covers and drip irrigation. The practice is costly and is only practical with fresh market vegetables. Benefits include: early harvest, increased early season yield, improved quality and reduced soil movement due to erosion. Removing and disposing of plastic materials in land fills after harvest is a drawback.

Insect-Pests and Diseases

Correctly identifying the insects and other pests that attack vegetables is the first step toward controlling these pests effectively. The pest descriptions in this booklet are designed to help the commercial vegetable grower and home gardener recognise the insects found on vegetables. Information on insect life cycles and habits is included to make pest control recommendations more understandable and usable. More specific descriptions of insects and detailed information on control procedures are available through local county Agricultural Extension Services offices.

Most of the insects considered common vegetable pests undergo a developmental process known as metamorphosis, which simply means that the insect changes form during its life. Metamorphosis may be complete or incomplete. Complete metamorphosis consists of four stages — egg, larva, pupa, and adult.

All insect life cycles include an egg stage. The insect is immobile during this stage. A larva hatches from the egg. The larval stage is the growth stage and frequently the pest stage of insect life. This is especially true of moths, butterflies, and true flies. Larvae of moths and butterflies are commonly known as caterpillars. Maggots are the larvae of true flies. Many beetles exist as grubs during the larval stage. Larvae vary in size from as short as 1/8 inch to as long as 4 inches.

The larva molts, or changes and enters the pupal stage. As during the egg stage, the insect is immobile during this pupal or changeover period. It is during this stage that the insect develops into an adult.

Adults may or may not feed in the same way and on the same plants as larvae. Beetles, for example, have chewing mouthparts and, like many grubs (beetle larvae), are capable of causing great damage to vegetables. Moths and butterflies, on the other hand, have mouthparts designed for siphoning and are unable to damage plants.

Incomplete metamophorsis consists of three developmental stages — egg, nymph, and adult. The egg stage is similar whether an insect undergoes complete or incomplete metamorphosis.

Immature insects called nymphs hatch from eggs. Nymphs damage plants and reduce vegetable quality by sucking plant juices. Some insects damage plants not only by extracting juice but also by injecting saliva into the plant. The best known example of this is the injury called hopperburn caused by the potato leafhopper on Irish potatoes.

The pests discussed in this booklet are divided into the following groups: Moths, Butterflies, and Their Young (Caterpillars); Beetles; Sap-Feeding Insects (True Bugs, Aphids, Leafhoppers, and Whiteflies); Flies; Other Insects; and Noninsect Pests.

Moths, Butterflies, and Their Young

Caterpillars, the larval stage of moths and butterflies, damage both the foliage and fruit of a number of vegetables. These insects chew holes in foliage and fruit and leave degrading excrement and silk on plants.

Cabbage Looper

This caterpillar feeds on the underside of leaves of cabbage and related crops and on lettuce, spinach, beets, peas, and tomatoes. It chews irregular holes through leaves. Moths lay greenish-white eggs on the upper surface of leaves. Larvae, or worms, are 1-1/2 inches long when full grown. Larvae are pale green with four thin, white lines along the back and a wide, pale line on each side of the body. These lines can be seen easily, even on small larvae. The head is narrower than the thorax (body sextion behind the head). The larvae move in a measuring or looping motion, thus the insect's name. When disturbed, loopers rear upwards. Larvae go into a resting or pupal stage in a flimsy, lace-like cocoon on the plant. The adult is a brown moth with silver markings on the wings. It has a wingspread of about 1-1/2 inches. There may be three or more generations per year, the number of worms increasing greatly with each generation. Cabbage loopers do not overwinter in North Carolina. Moths from Georgia and Florida move into the state each year near the middle of May.

Beet Armyworm

Similar to the true armyworm, this insect is a general feeder and attacks foliage, stems, and sometimes roots of a wide variety of vegetables. Among plants attacked by this pest are beets, asparagus, corn, lettuce, cabbage, tomatoes, peppers, potatoes, onions, and peas. This pest is capable of destroying young plants. The beet armyworm overwinters as a pupa, emerging as an adult in the spring. The adult is a moth with a 1-1/4 inch wingspread. The moth's grayish brown front wings are marked with lighter lines and a spot near the centre of the wing. Hind wings are white with brown edges and veins. Females lay between 500 and 600 eggs over a 4- to 10-day period. Irregular egg masses are covered with scales from the moth. Mature larvae are green with a prominent dark stripe along each side. Larvae have a dark spot on each side of the thorax, are 1-1/4 inches long, and pupate in the soil. In North Carolina, there are several generations a year.

Corn Earworm

This caterpillar feeds on a wide range of vegetables and, as a result, is known by several names. It may attack almost any vegetable but does the most serious damage to corn and tomatoes. Other vegetables attacked include peppers, eggplant, beans, okra, sweet potatoes, lettuce, and cabbage.

Fresh corn silk is preferred by this pest as a location to deposit eggs. Eggs may hatch as soon as two or three days from the time they were deposited. Newly hatched larvae feed on buds and leaves or on corn silks. When they are four or five days old, larvae travel down the silks and feed on kernels at the end of the corn ear. If fresh corn silk is unavailable, moths prefer laying their eggs on tomato foliage, especially when plants are flowering. Larvae attack tomatoes from the time the fruit forms until it ripens, boring into the sides of the fruit and near the stem. Sometimes a small, black entrance hole is the only visible sign that a large worm is inside a tomato. Injury to tomatoes is most severe in the late summer and fall.

When full-grown, larvae are about 1-1/2 inches long. They may be brown, green, or pink, with lighter coloured longitudinal stripes. The head is yellow, and the underside of the body is always lighter than the back. Larvae enter the soil to pupate.

Moths are cream to grayish-brown, have green eyes, and are marked with a dark spot near each wing tip. They have a wingspread of about 1-1/2 inches. Hind wings are light coloured with darker areas near the edges. Moths deposit yellow eggs one at a time on plants. One female can lay as many as 3,000 eggs. There are usually three and sometimes a partial fourth generation of earworms each year in North Carolina.

Cutworm

Many kinds of cutworms attack vegetables. Asparagus, cabbage, squash, and tomatoes are particularly vulnerable. Most cutworms cut off stems of plants at or near the soil. Cutworms feed at night when temperatures are 65 degrees F or higher. During the day these insects hide just beneath the soil close to the site of the previous night's damage. This pest is capable of damage sufficient to necessitate the replanting of corn. Cutworms usually do not eat plants beyond cutting through them, although some species climb stalks and feed on the upper parts of plants, which causes less damage.

When full grown, cutworms are 1-1/2 inches long. They curl up into a tight C shape when disturbed. Most cutworms are a dull colour and have practically no hair. They pupate underground. Cutworms generally overwinter in the soil in the larval or pupal stage. Adults are dull coloured moths with wingspreads of 1 to 1-3/4 inches. Hind wings are usually light in colour. Moths lay eggs on the stems of grasses and weeds or on bare ground. There may be more than one generation of some species each year.

Diamondback Moth

Larvae of this moth feed during the cool parts of the growing season on the underside of leaves of cabbage and related crops. Larvae feed for a time after hatching as leafminers. As larvae grow, they begin feeding on the external portion of leaves and buds. Feeding holes do not penetrate the entire leaf. Larvae are greenish-yellow with black hair and are about 1/3 inch long when full grown. They wriggle actively when disturbed and drop from the plant on silken threads. Larvae pupate inside a lacy cocoon attached to the leaf of the host plant. Moths overwinter in debris of collards, cabbage, cauliflower, and related crops. They have a wingspread of about 1/3 inch. Moths are gray to brown, marked with white. When at rest the moth has white, diamond-shaped markings along the back. The hind wings are paler and have a fringe of hair on the rear margin. Moths lay small, white eggs either singly or in groups of three or four on leaves. There may be five or six generations per year in North Carolina.

European Corn Borer

This borer attacks many species of plants, but corn, peppers, and Irish potatoes are the vegetables most severely damaged. European corn borers feed on leaves and in protected areas until about half-grown, when they bore into a plant's stem or fruit. Nearly an inch long when full grown, the caterpillar is cream coloured, has a brown head, and is marked with small, round, brown spots. Caterpillars molt into pupae inside the host plant. This insect overwinters as a nearly full-grown larvae in cornstalks at ground level. Adults are pale yellow moths with irregular dark bands running across the wings. The moth's wingspread is about 1 inch. Males are smaller and somewhat darker than females. Eggs are laid in groups of up to 50 on the underside of leaves and overlap to resemble small fish scales. Most of North Carolina sees at least two generations of this insect each year, while in the eastern part of the state, there are three generations most years.

Fall Armyworm

Like other armyworms, this insect may "march" from field to field. It often is called the shatterworm because of the destruction it causes to corn foliage. Larvae also damage corn ears as well as beans, potatoes, peppers, tomatoes, spinach, cabbage, and cucumbers. Swarms of moths move northward each spring from southern Florida and areas along the Gulf Coast. Females lay about 1,000 eggs each in masses of about 150. Eggs are usually laid on grass. Larvae are very small when they hatch but reach a length of 1-1/4 inches when full grown.

Larvae vary in colour from light tan to green or nearly black, are decorated with thin, yellow stripes, and have a prominent inverted white Y on the front of the head. Larvae go into the soil to pupate. Adult moths have dark gray front wings that are mottled with light and dark splotches. The wingspread of the adult moth is about 1-1/4 inches. Hind wings are grayish-white. Males have a white spot near each wing tip. It appears that this insect cannot overwinter in North Carolina. Most years there are many generations of the pest.

Imported Cabbageworm

This caterpillar attacks cabbage, cauliflower, kale, collards, mustard, radish, and turnips. Larvae riddle outer leaves with irregularly shaped, often large holes. Caterpillars are capable of doing enough damage to severely retard plant growth. Females lay one egg per leaf on the underside of leaves. Eggs hatch into caterpillars that are 1-1/4 inches long when full grown. Caterpillars are velvety-green and are decorated with three indistinct gold stripes, one along the back and one on each side. The pupal or chrysalid stage of the insect is brownish-grey and has many projections or spines. This insect is commonly found attached to a leaf during the pupal stage. Glue at the pupa's tail and a silken loop that encircles the middle of the pupa's body hold it in place. The adult is a white butterfly with one or two black spots on each wing. This pest overwinters as a pupa either on old plant refuse or on nearby posts or buildings. There are three or four generations per year.

Pickleworm

Pickleworm larvae bore into many cucurbit fruits. Cantaloupes, cucumbers, and squash are most seriously damaged. Caterpillars feed first on foliage. As the insect grows, it begins feeding on fruit. Pickleworms do the greatest damage to growing vine tips and blossoms early in the season. Larvae are initially white with black spots and have dark heads. There is a dark indention on each side on the back of the head. Larger larvae are up to 3/4 inch long and are green. Sawdust-like excrement produced by larvae can easily be seen extruding from damaged fruit. Winter is spent in the pupal stage in a silken cocoon on crop refuse in subtropical regions of the United States. Pickleworms usually appear in North Carolina in early July. The adult is a moth with a wingspread of about 1 inch. Wings appear fragile and are marked with pale yellow centres and yellowish-brown margins.

Potato Tuberworm

This caterpillar does significant damage feeding on exposed potatoes in the field, riddling them with narrow, silk-lined burrows. Adults are small, narrow-winged, grayish-brown moths with a wingspread of 1/2 inch. Moths tend to fly in late afternoon and evening. Females deposit single eggs on foliage of tomatoes, potatoes and related plants, or on exposed tubers of Irish potatoes. Eggs hatch into pinkish-white or green caterpillars with brown heads. When full grown, caterpillars are about 1/2 inch long. Larvae mature in about three weeks and pupate in gray, silken dirt-covered cocoons on the outside of the tuber. A brush of long dark scales protrudes at the end of the body. Moths deposit clusters of two to seven eggs on fruit, vines, buds, and leaves. There are at least four generations per year in North Carolina. Cultural control is possible by planting for early maturity and destroying plants immediately after harvest, thus preventing movement of the pest to later crops.

To prevent damage, potatoes should be kept covered with soil before harvest, and harvested potatoes should be stored in insect-proof areas. Also, vines should be cut, raked, and burned two or three days before harvesting. There may be several generations of this insect per year.

Squash Vine Borer

This borer can be very destructive to squash and pumpkin plants. By boring into a vine near the base, this thick, white, wrinkled brown-headed caterpillar causes entire plants to wilt. The larva overwinters 1 or 2 inches in the soil within a dark, dirt-encased cocoon and molts into a pupa in the spring. Within two or three weeks after molting, the 3/4-inch-long pupa has worked its way to the soil surface and emerged as an adult. The adult squash vine borer is a moth with a wingspread of about 1-1/2 inches. The front wings are covered with metallic, greenish black scales.

Hind wings are transparent. The abdomen is ringed with red and black. Squash vine borer moths look and act like paper wasps. Females lay red eggs near the base of plant stems. Eggs produce smooth, white caterpillars that bore into the centre of the vine a few days after hatching. Small piles of green excrement and wilting of whole vines are signs a plant is infested with squash vine borers. There are two generations of this pest each year. Damage is usually noticed in early June and again in early August.

Tomato Hornworm and Tobacco Hornworm

Tomato and tobacco hornworms feed on tomatoes, peppers, potatoes, and related plants. These green worms are 3 to 4 inches long and can defoliate young plants. Mature tomato hornworms can be distinguished by eight L-shaped, white markings along each side of the body and a bluish-black horn near the rear end of the insect. Tobacco hornworms are decorated with diagonal white bars and have a red horn. The white cocoons of a parasite are usually attached to the worm. Pupae of both species are mahogany brown and are found in the soil. In both species, adults are large, gray moths with mottled front wings. Hind wings are lighter. Two thin, zigzag lines run across the midsection of each hind wing of tomato hornworm moths. There is a dark area between these somewhat obscure lines on the wings of tobacco hornworm moths.

On both species, five or six yellowish-orange spots can be found along each side of the body. Moths of both species have a wingspread of 4 to 5 inches and fly mainly around dusk. Females lay eggs individually on plants. Eggs are spherical and greenish-white. This insect overwinters as a pupa. In North Carolina, there are two generations a year.

True Armyworm

This caterpillar is sometimes present in very high numbers. Armyworm epidemics can result in the destruction of large areas of vegetation. This pest attacks grasses, corn, and some legumes. Damage usually starts at the edge of a field. Larvae feed at night and may move in hordes. Armyworms are capable of consuming corn plants up to 8 inches tall. This insect stops feeding and overwinters as a partially grown larva. It resumes feeding in the spring. Armyworms are dark greenish-brown worms up to 2 inches long with white stripes down the sides and back. The insect pupates just below the surface of the soil.

Yellowstriped Armyworm

This caterpillar is a daytime feeder. It is capable of doing significant damage to young tomato plantings, pepper fruit, and, occasionally in August, to sweet potatoes. Its presence is sporadic, however. Overwintering as pupae in the soil, yellowstriped armyworms emerge as moths in the spring. Adults are brown, black, and white moths with a 1-1/2 inch wingspread. The moth's forewings are marked with a diagonal band that runs from the front to the rear of the wing and a smoky white stripe along the outside edge of the wing. Hind wings are white with a dark edge. Eggs, covered with scales from the female's body, are laid in clusters on foliage, trees, and buildings. Larvae are dark green with a yellowish-orange stripe along each side and black spots on their abdominal segments. In North Carolina there are several generations a year.

Beetles

Adult beetles are usually hard-bodied insects with thick forewings. The young are grubs, borers, or wireworms. Often adults feed on different host plants than do larvae, although both stages may be destructive to vegetables.

Asparagus Beetle

There are two beetle species that feed on asparagus — the asparagus beetle and the spotted asparagus beetle. The spotted asparagus beetle, which is reddish-orange and black, is more destructive in northern asparagus production areas. Asparagus beetles are more troublesome in North Carolina. The asparagus beetle is metallic blue with six yellow areas on the wing covers. Beetles are about 1/4 inch long and have a narrow, red head and thorax. This pest overwinters in protected areas and first appears each year in early April along the edges of fields. Adults gnaw tender asparagus buds and lay eggs on developing spears. Eggs are very dark, bulletlike, and laid in rows. Processing plants sometimes reject otherwise edible spears because eggs have been laid on them. Eggs hatch within a week of the time they were laid. Sluggish, humpbacked, slate-coloured larvae with black heads feed and grow on asparagus ferns for about 10 days. The pupal stage is passed on the soil. A week or so after pupation, adults emerge and the cycle is repeated. There may be as many as five generations in a year. Asparagus is the only vegetable this insect attacks.

Bean Leaf Beetle

Adult bean leaf beetles range in colour from red to yellow. The insect's markings may also vary, but bean leaf beetles generally have a dark triangular mark on the front of the wing covers, six black spots in the centre of the back and a black edge around the outside of the wing covers. This insect is about 1/4 inch long and is long-lived. Adults will drop from foliage when disturbed. Bean leaf beetles feed on foliage, making round, regular holes. This pest also attacks the stems of young plants at or slightly below ground level. Females lay eggs in the soil near the base of plants. From the eggs hatch slender, white larvae that feed on the roots of bean plants. Bean leaf beetles spend the winter as adults in trash near bean fields. There is one generation, possibly two, per year. Beans, peas, cowpeas, and soybeans are common host plants; however, this pest will also feed on corn, some clovers, and many weeds.

Blister Beetle

Several species of blister beetle feed on vegetables. The black, striped, and margined blister beetles are most commonly seen in North Carolina. Blister beetles are slender, rather soft insects with a narrow thorax. The tip of the abdomen is exposed beyond the wing covers. Adults are black or gray and sometimes have narrow, yellow or gray stripes or edges on the wings. Adult beetles are ravenous feeders and are capable of destroying plants. The winter is passed as larvae in the soil. Larvae are about 1/3 inch long and yellow. Blister beetle larvae are beneficial in that they feed on grasshopper eggs. The pupal stage also is spent in the soil. Adults emerge from the soil in midsummer and feed in groups for about five weeks. The beetle's body contains an oil called cantharidin that causes blisters if the insect is crushed against tender skin. Yellow eggs are laid in holes in the soil in clusters of 100 or more. It appears there is one generation per year. Blister beetles will attack most types of vegetables but seem to prefer potatoes and tomatoes.

Colorado Potato Beetle

This insect is perhaps best known as the common yellow- and black-striped potato bug. This pest is, however, a beetle and not a true bug. Both adults and larvae feed on the foliage of potatoes. While potatoes are preferred, this pest also attacks tomatoes, eggplant, peppers, and horse nettle. Adults are about 3/8 inch long and 1/4 inch wide and very convex in shape. Black and yellow stripes run lengthwise along the insect's back. Females lay clusters of 20 or more yellowish-organe eggs on the underside of leaves. A single female beetle can lay over 500 eggs over the course of her four-week life. From the eggs come red, hump-backed, soft-bodied larvae decorated with two rows of black spots along each side of the body. Larvae begin chewing on leaves immediately after hatching. Larvae also feed on the fruit of tomatoes and eggplant. In two or three weeks larvae grow to a length of 1/2 inch, then enter the soil to pupate. Seven days later adult beetles emerge from the soil, feed for a few days, mate, and lay more eggs. This insect overwinters as an adult in the soil. There are at least two generations per year.

Cowpea Curculio

This insect most commonly attacks blackeyed and crowder peas. Adults are black, humpbacked snout beetles about 1/4 inch long. The adult female chews a hole in a pea pod and inserts an egg into the developing seed. This small puncture is difficult to see, and damage may go unnoticed until the crop is harvested. Eggs hatch into small, white, legless grubs with yellowish-brown heads. The grubs destroy the peas in a pod. There is at least one generation per year. Hot, dry weather appears to retard development of this pest. Adults overwinter in refuse around fields. Damage may be avoided by alternating from year to year the site where peas are planted.

Flea Beetle

The many different species of flea beetle attack a variety of vegetables. Species that attack potatoes, tomatoes, and eggplant are the most troublesome, but flea beetles also damage corn and sweet potatoes. Flea beetles are so named because of their enlarged hind legs and jumping ability. Adults are 1/16 to 1/18 inch long and usually black with brown legs and antennae. Eggs are laid in the soil at the base of plants. From the eggs hatch cylindrical, brown-headed, white larvae that are about 1/5 inch long when full grown. Larvae commonly feed on roots, often riddling them with tunnels or eating off rootlets. The pupal stage is spent in the soil. These insects overwinter as adults in trash around field borders and in ditch banks. There are three or more generations per year. Round holes chewed in leaves are the most obvious damage caused by flea beetles. On small plants, this "buckshot" damage can result in plant death. On larger plants these feeding sites provide a good entrance for several blights and wilts.

Japanese Beetle

These beetles are metallic green or greenish-bronze, 1/2 inch long with red wing covers, and are decorated with white spots near the tip of the abdomen. They commonly are found in June and July on fruit, deciduous trees, and vegetables. Japanese beetles also are frequent and persistent pests of sweet corn and beans. The insect spends the winter in the soil as a white grub about 3/4 inch long. Beetles emerge from the soil in midsummer and on warm, sunny days swarm to plants, where they eat the foliage, flowers, and fruit. Females beetles then return to grassy areas to lay their white, spherical eggs in the soil. Grubs feed on decaying vegetation and grass roots. Pupation occurs in the soil. There is only one generation each year.

Mexican Bean Beetle

The brown Mexican bean beetle is decorated with 16 black spots arranged in three rows across its back. Beetles are 1/3 inch long, convex, and resemble lady beetles. Adults overwinter on the ground among leaves and trash near fields where beans were grown. When warm weather arrives in the spring, they leave their overwintering sites and feed on beans for a week or two. Females lay lemon-coloured eggs in groups of 25 or more on the underside of leaves. Eggs hatch about 10 days after being laid and yellow larvae with six rows of black-tipped, branched spines on their backs appear. Larvae, which are 1/3 inch long when full grown, feed for two to five weeks, then enter the pupal stage. The yellow pupa is attached to a leaf. Traces of the last larval skin can usually be found where the pupa is attached to the leaf. There are three or four generations per year. Both larvae and adults feed on leaves, usually on the underside. Damaged leaves have a skeletonized or lace-like appearance. Sometimes stems and pods are damaged. The insect attacks all kinds of garden beans, cowpeas, and soybeans.

Spotted Cucumber Beetles

In its adult form, this insect feeds on over 200 different plants. It may be found on beans, cucurbits, corn, potatoes, asparaugs, cabbage, peas, beets, tomatoes, and turnips. It feeds on foliage and new shoots, leaving small holes in foliage. It also gnaws holes in fruit. This insect overwinters as a 1/4 inch long, yellowish-green beetle decorated with 12 black spots. The beetle's head and antennae are black. Females deposit eggs at the base of grass and corn plants. From the eggs hatch yellowish-white, brown-headed larvae that are up to 3/4 inch long when full grown. These larvae, called corn rootworms, tunnel in the roots of plants. This tunneling action can result in severe stunting of corn plants and sometimes kills plants. There appears to be two generations per year in North Carolina.

Striped Cucumber Beetle

These 1/5-inch long beetles are yellow and are decorated with three wide black stripes. Striped cucumber beetles damage plants by feeding on leaves, stems, and fruit. They also carry the organisms that cause bacterial wilt of cucurbits and cucumber mosaic. Adults lay yellow eggs at the base of plants. Larvae, which are white and 1/3 inch long, feed on the underground parts of plants or on fruit that touches the soil. There are at least two generations per year. In addition to cucurbits, striped cucumber beetles feed on beans, peas, and corn. The most severe damage is caused to small plants.

Vegetable Weevil

Larvae and adults of this insect feed on foliage and exposed roots of potatoes, turnips, turnip greens, kale, and mustard greens. Late fall and early spring crops are most severely damaged. Adults are gray snout beetles 1/3 to 1/2 inch long. Two white markings on the rear half of the wing covers form a V shape. Eggs laid in the crowns of plants procude green, slug-like larvae. These grubs are about 1/2 inch long when full grown. This pest goes into a resting stage from June to September.

Wireworms

These beetles do the most severe damage to root crops such as Irish and sweet potatoes, although they injure other vegetables as well. There are many different species of wireworms. Adults are hard-shelled, dull coloured beetles, ranging from 1/2 to 1-1/2 inches in length. Beetles are tapered toward both ends and appear streamlined. Adults are also known as click beetles because they snap and flip over when turned on their backs. They have a loose, flexible joint just ahead of the wings. Females lay eggs in the soil around the roots of grass plants. The eggs produce yellow to reddish-brown, tough-bodied larvae that are commonly called wireworms. Larvae range in size from 1/2 to 2 inches when full grown, and most are cylindrical and slender. Some wireworms are almost flat. Larvae feed on underground parts of plants. The pupal stage is spent in the soil. It takes over a year to complete the life cycle of a wireworm. Wireworm damage is most severe on land not previously planted in row crops.

White Grubs

There are several species of these pests. Adults are hard-shelled, light brown to almost black beetles from 3/4 to 1 inch long. These insects are sometimes called May beetles and June beetles. They commonly feed at night on the foliage of trees and hide in the grass and lay eggs during the day. The spherical, pearly white eggs produce white C-shaped grubs with brown heads. Grubs feed on grass roots and other crops. The pupal stage is spent in the soil. Among the common white grubs are the larvae of the green June beetle. Larvae feed mainly on organic matter in the soil. Grubs uproot young plants as they move about in and on the soil. Vegetables should not be planted in soil that was in sod the previous year until the soil has been treated to control white grubs as some grubs live two to three years.

Whitefringed Beetle

Among the many vegetables attacked by whitefringed beetles are Irish potatoes, sweet potatoes, beans, and cabbage. Adults, which emerge in June and July, are dark gray beetles about 1/2 inch long, with a white margin or fringe on their sides. White, oval eggs are laid in a gelatinous mass on objects on or near the ground. Grubs are shite, legless, and 1/2 inch long when full grown. Grubs feed on the underground parts of many kinds of plants. The pupal stage is spent in the soil, while the insect overwinters in the grub stage. There is one generation per year.

Sap-Feeding Insects

Some insects damage vegetables by sucking plant juices from leaves, stems, or roots. The true bugs, aphids, whiteflies, and leafhoppers, all have beak-like mouthparts that are used to suck sap from plants. Some of these insects also produce toxic saliva, which they inject into plants. The best known example of plant injury caused by saliva injection is a condition called hopperburn caused by the potato leafhopper on Irish potatoes. Many of these insects also transmit plant diseases. The green peach aphid, for example, is known to transmit potato leafroll virus, while the melon aphid infects squash and watermelon mosaic virus.

Harlequin Bug

These bugs feed on cabbage and related crops and can cause severe damage. Harlequin bugs also feed on beans, asparagus, okra, and tomatoes. The insect damages a plant by extracting juices from it. Adults and nymphs have similar habits. Adults are flat, shield-shaped bugs about 3/8 inch long. They are red or organe with black markings that give the appearance of a mask. Eggs resemble kegs and are laid on end in groups. Each egg has a black band at the top and bottom. Nymphs, which grow through five developmental stages, resemble adults but are smaller and wingless. There are two or more generations of this insect per year. Adults overwinter in any protected area.

Eggplant Lace Bug

Adult eggplant lace bugs are small, flat insects about 1/8 inch long. They are somewhat rectangularly shaped when viewed from above. The wings are net or lace-like. The head is covered with a hood-like structure. Females lay eggs on the underside of leaves. Nymphs are darker than the adults, wingless, and usually covered with spines. Both nymphs and adults inject saliva and suck juices from eggplant leaves. The insect also excretes a brown, varnish-like material.

Squash Bug

This insect attacks only cucurbits. It prefers squash and pumpkins. Adults and nymphs injure plants by sucking sap from them. Damage usually can be detected by the occasional drooping of a leaf. Adults are about 5/8 inch long, brownish-black or gray, and flat across the back. This insect overwinters as an adult and mates in the spring. Reddish-brown eggs are laid on leaves in rows that come together at an angle. Each cluster contains 15 or more eggs, which produce tiny, light gray, wingless nymphs. There is one generation each year. Destruction of crop residue as soon as the crop is harvested helps to control these insects.

Tarnished Plant Bug

Among the many different vegetables this insect attacks are asparagus, Irish potatoes, chard, celery, and beans. A small spot surrounded by a yellow area is evidence a tarnished plant bug has been feeding on a bean seed. Blossoms may drop from a plant as a result of tarnished plant bugs feeding on them. This insect also feeds on the terminals of potatoes, causing leaves near the top of the plant to wilt. Tarnished plant bugs also transmit disease. Adults are small, flat bugs about 1/4 inch long. These somewhat oval insects are mottled brown with irregular splotches of white, yellow, reddish-brown, and black. Eggs are usually inserted into plant tissues. The nymphs resemble adults but are smaller and wingless. There appear to be four or five generations per year in North Carolina. This insect overwinters as an adult. Two other true bugs — stink bugs and leaffotted bugs — cause similar injury, especially on beans and southern peas. Damage inflicted by these two bugs on southern peas and butter beans occasionally is serious.

Aphids

Aphids, or plant lice, feed on most vegetables. These insects cause the greatest damage when they suck juices from and inject saliva into plants. Many aphid species transmit viruses that cause diseases. Adults are soft-bodied, pear-shaped, and may or may not have wings. Two cornicles, or "exhaust pipes," extend from the hind end of most aphids. Aphids may be green, pink, yellow, or black. Nymphs resembe adults but are smaller and always wingless. These pests usually feed in clusters.

Potato Leafhopper

This insect feeds on many different host plants including beans and Irish potatoes. Saliva injected by this pest into Irish potato plants causes a deadening and upcurling of the tissues known as hopperburn. Sap feeding on beans causes leaves to be stunted, crinkles, and curled downward. Adults are pale green, wedge-shaped, and about 1/8 inch long. Eggs inserted into plant tissues produce green nymphs that resemble adults but are smaller and wingless. Adults and nymphs suck plant juices. Potato leafhoppers are a problem primarily in western North Carolina. At least three generations of this pest occur each year. The potato leafhopper is one of several leafhopper species found on vegetables.

Whitefly

Whiteflies feed on a wide variety of vegetables. This insect can be found in the field and the greenhouse. Host plants include beans, tomatoes, eggplant, and cucumber. Adults are very small sucking insects with two pairs of broadly rounded wings. The wings and body are covered with a snow white, waxy powder. The insects look like tiny moths and may fly up in a cloud when disturbed. Eggs, which are almost microscopic, are laid on the underside of leaves. Young nymphs are small, flat, and hard to see. All stages are found on the underside of leaves.

Leaves may become discoloured and fall from a plant on which whiteflies are feeding. Plants on which this insect feeds may also be stunted. Whiteflies secrete honeydew on plants, which allows unsightly black sooty mold to grow.

Flies

True flies are a diverse group. Many species attack humans and animals and can spread diseases. Some flies are beneficial. They feed on decaying matter and prey on or parasitize other insects. Others are plant pests, their maggots or larvae feeding on various vegetables.

Cabbage Maggot

This insect prefers cabbage, cauliflower, broccoli, and related crops. It also attacks beets, radishes, turnips, celery, and other vegetables. The cabbage maggot does most of its damage in the mountain regions of North Carolina. The adult is a gray fly with two wings. It is smaller than a house fly but has proportionately longer legs. Eggs are white and are laid on the soil close to young plants. The eggs hatch into white, wedge-shaped maggots that are about 1/4 inch long when full grown.

Maggots feed on roots, either on the surface or by tunneling through them. Fleshy parts of radishes or turnips will show brown streaks where tunneling has occurred. Cabbage maggots overwinter in the pupal stage, which is spent in the soil. There appear to be three generations per year of this pest.

Pepper Maggot

This pest prefers peppers but also attacks eggplant. It is believed that its natural food is horsenettle. Adults are yellow striped flies about 1/4 inch long with dark bars on the wings. Females deposit eggs under the skin of peppers.

Maggots are white, legless, and about 1/2 inch long when full grown. Maggots eat the core of the pepper, causing it to decay or drop from the plant. The pupal stage, in which the insect overwinters, is spent buried 2 to 4 inches in the soil. Adult flies appear in late June. There is one generation each year.

Seedcorn Maggot

This insect attacks the sprouting seeds of many different vegetables. Damage may be done to seed pieces of Irish potatoes; seedlings of corn, beans, peas, and cucumbers; or stems of other vegetables. The most serious infestations occur in the spring during cool, wet weather and in soils high in organic matter. Adults are grayish-brown flies about 1/5 inch long. Eggs, from which hatch cream-coloured, wedge-shaped maggots, are deposited in the soil close to plants. Maggots are 1/4 inch long when full grown, legless, and have chewing mouthparts at the narrow end of the body. It is as a maggot that this insect is destructive. This pest overwinters in the soil in the pupal stage. The pupa is enclosed in a brown, capsule-like case. There are three generations of this pest each year. Destruction of crop refuse and shallow planting in warm soils will enhance plant establishment.

Vegetable Leafminer

This insect's name derives from the maggot's habit of tunneling or mining in leaves — feeding between the upper and lower leaf surfaces. Damage appears as winding white trails or broad white spots. Leafminers attack many vegetables but do the most damage to cucumbers, squash, tomatoes, and leafy vegetables such as cabbage, kale, and collards. The leaves of a plant that is attacked by vegetable leafminers are weakened, while the mines serve as sites for disease and decay.

Adults are gray to shiny black flies less than 1/8 inch long. Small, white eggs deposited in leaves produce white or yellow maggots that tunnel through the leaves. The tunnel is narrow and winding at first. It becomes larger as the maggot grows, finally forming a blotch. The pupal stage is spent in the soil, and it is in this stage that this pest overwinters.

Noninsect Vegetable Pests

Spider mites and slugs, although not insects, are capable of inflicting severe damage on vegetables.

Spider Mites or Red Spiders

Mites are closely related to insects. Many species of this pest attack vegetables. Beans and tomatoes may be severely damaged, while crops such as eggplant, cucumber, and celery are also susceptible. These pests appear to cause more severe damage during hot, dry years. Adults are usually red, yellow, or green, about 1/ 50 inch long, and have four pairs of legs. The body is oval. Adults lay eggs on plants. The eggs hatch into young mites that resemble adults but are smaller. Adults and nymphs suck plant juices. Plants attacked by mites appear sickly and white-speckled. Affected leaves curl, turn brown, and drop from the plant prematurely. When an infestation is severe, fine silk webs may entangle an entire plant, and moving spider mites can easily be seen in the webbing. Many generations may occur each year.

Slugs

Slugs are similar to snails but lack the snail's hard, protective shell. Six species of slugs attack vegetables and ornamental plants in North Carolina. Slugs feed on mold, decaying organic matter, and live plant tissue including foliage, flowers, or fruit. Slugs may chew jagged edges along leaves, shred leaves, or eat completely through leaves or stems. Damage is most severe in damp areas or during extended rainy periods. Slugs vary in size from less than 1/2 inch to more than 4 inches long and range in colour from ash-gray or yellow-gray to black. Some are mottled. Young slugs resemble adults but are smaller and lighter in colour. Eggs, gelatinous and water appearing, are laid in clusters of about 25 on the soil in concealed, moist areas. Slugs are active at night, leaving silvery slime trails that can be seen the next morning.

Bibliography

Ashraf, Syed Mahboob: *Handbook of Fruit and Vegetable Products*, Agrobios, Delhi, 2008.

Bhutani, R.C.: *Fruit and Vegetable Preservation*, Biotech Books, Delhi, 2003.

Chambers, R.: *Rural Development: Putting the Last First*, London, Longman, 1983.

Collymore L.: *Fruit Production in Barbados*, Port of Spain, Trinidad and Tobago, 1996.

Coste R.: *Coffee: the Plant and the Product*, London, Mac Millan, 1992.

Crucefix D.: *Avocado Variety Selection for Export Development*, Roseau, CARDI, 1996.

Doijode S. D.: *Seed Germination in Fruits*, New Delhi, Malhotra Publishers, 1993.

Ferentinos L.: *Proceeding of the Sustainable Taro Culture for the Pacific Conference*, Honolulu, HITAHR, 1993.

Gupta, S.K.: *Disease Problems in Vegetable Production*, Scientific, Delhi, 2006.

Hessayon D. G. Dr.: *The Vegetable Expert*, England, PBI, Publications, 1985.

James S Shoemaker: *Culture of Vegetable Growing*, Asiatic Pub, Delhi, 2006.

Jeffers P.: *Evaluation of Four Onion Varieties in Montserrat*, Plymouth, CARDI, 1992.

Julia F.: *Fruits of Warm Climates*, Miami, Julia F. Morton Publisher, 1987.

Kataria, T N: *Plant and Crop Physiology*, Pearl Books, Delhi, 2008.

Kumar, N.: *Breeding of Horticultural Crops : Principles and Practices*, New India Pub, Delhi, 2006.

Mellor, J. W.: *The New Economics Growth of Vegetable Production*, Ithaca, Cornell University Press, 1976.

Morris M. L. : *Maize Seed Industries in Developing Countries*, Colorado, Lynne Rienner, 1998.

Narang, R.K.: *Fruit and Vegetable Preservation Techniques*, APH Pub, Delhi, 2010.

Oldham P.: *Cost of Production of Major Tree Crops in Dominica*, Roseau, Ministry of Agriculture, 1991.

Rana, M. K.: *Breeding and Protection of Vegetables*, New India Pub, Delhi, 2011.

Rao , P. Venkateshwara: *Dairy Farm Business Management*, Biotech Books, Delhi, 2008.

Rehm Sigmund: *Multilingual Dictionary of Agronomic Plants*, Boston, Kluwer Academic Publishers, 1994.

Seshadri, V.S.: *Cucurbit Vegetables : Biology, Production and Utilization*, Studium Press, Delhi, 2009.

Singh, Kirti: *Emerging Scenario in Vegetable Research and Development*, Research Periodicals and Book Publishing House, Delhi, 2001.

Singh, Neeraj Pratap: *Basic Concepts of Vegetable Science*, International Book, Delhi, 2004.

Singh, S.K.: *Biotechnology, Plant Propagation and Plant Breeding*, Campus Books, Delhi, 2008.

Somasundaram Rajarathnam: *Advances in Preservation and Processing Technologies of Fruits and Vegetables*, New India Publishing Agency, Delhi, 2011.

Srivastava, A.P.: *Mechanization of Vegetable Production and Post-Harvest Management*, Westville Pub, Delhi, 2009.

Steferud, Alfred: *Diseases of Fruits and Nuts*, Biotech Books, Delhi, 2005.

Thomas E.: *Fruit Production in St. Kitts and Nevis*, Port of Spain, IICA, 1996.

Tripathi, Pramila: *Botanical Pesticides in the Management of Post-harvest Fruit Diseases*, Daya, Delhi, 2005.

Verma, L.R. and Sharma, R.C.: *Diseases of Horticultural Crops: Fruits*, Indus, Delhi, 1999.

Walia, S. and B.S. Parmar: *Pesticides, Crop Protection and Environment*, Oxford & IBH, Delhi, 1995.

Yadav, D.S.: *Advances in Vegetable Production*, Researchco Book Centre, Delhi, 2005.

Index

❑❑❑